Nuclear Materials Science

Nuclear Materials Science

Karl Whittle
School of Engineering, University of Liverpool, UK

IOP Publishing, Bristol, UK

Media content for this book is available from http://iopscience.iop.org/book/978-0-7503-1103-8/page/about-the-book.

ISBN 978-0-7503-1104-5 (ebook)
ISBN 978-0-7503-1105-2 (print)
ISBN 978-0-7503-1128-1 (mobi)

DOI 10.1088/978-0-7503-1104-5

Version: 20160501

IOP Expanding Physics
ISSN 2053-2563 (online)
ISSN 2054-7315 (print)

British Library Cataloguing-in-Publication Data: A catalogue record for this book is available from the British Library.

Published by IOP Publishing, wholly owned by The Institute of Physics, London

IOP Publishing, Temple Circus, Temple Way, Bristol, BS1 6HG, UK

US Office: IOP Publishing, Inc., 190 North Independence Mall West, Suite 601, Philadelphia, PA 19106, USA

For those who may find this useful.

Contents

Preface

There are 442 reactors operating in the world, generating a total of ~383 GW of electricity, with another 66 in various stages of construction, as shown in figure 0.1. There are various arguments both for and against nuclear power, ranging from the view that it should not be used because the waste is harmful through to the belief that it is necessary to generate the large amount energy needed for modern society to function.

Nuclear reactors are often considered to be highly complex and technical beasts, and in many ways they are. However, the principle behind a nuclear reactor is essentially the same as for most other power stations, i.e. the conversion of heat to electricity through a turbine, predominantly a steam turbine, as shown in figure 0.2.

Fundamentally, therefore, a nuclear reactor can be compared to a steam engine, with the reactor core as the firebox, boiling water to make steam, which is then used to drive the turbine. In essence, this the nature of a nuclear reactor, it is how we

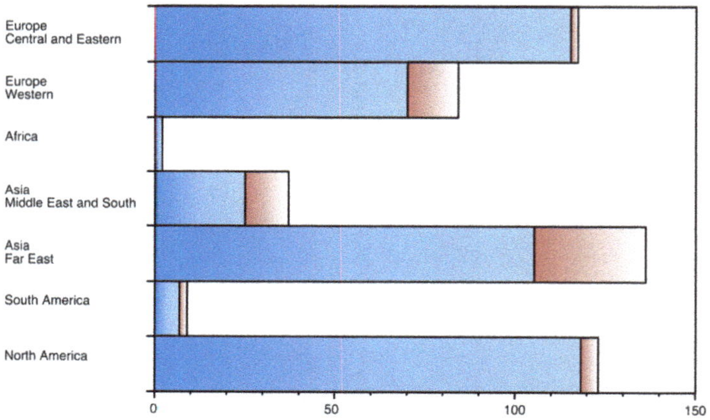

Figure 0.1. Outline of reactors operating and under construction worldwide. Data taken from [1].

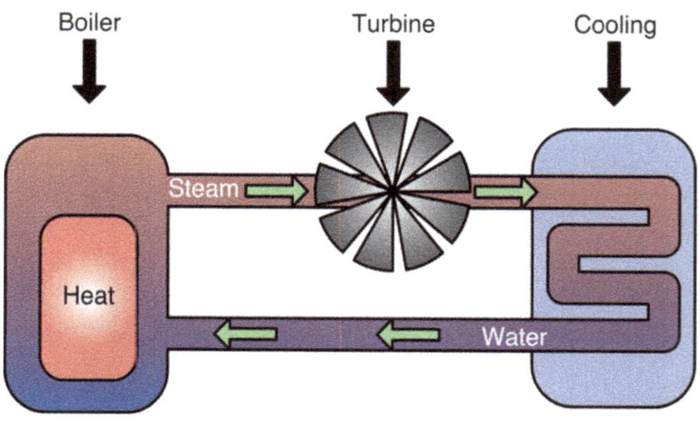

Figure 0.2. Schematic of a power station.

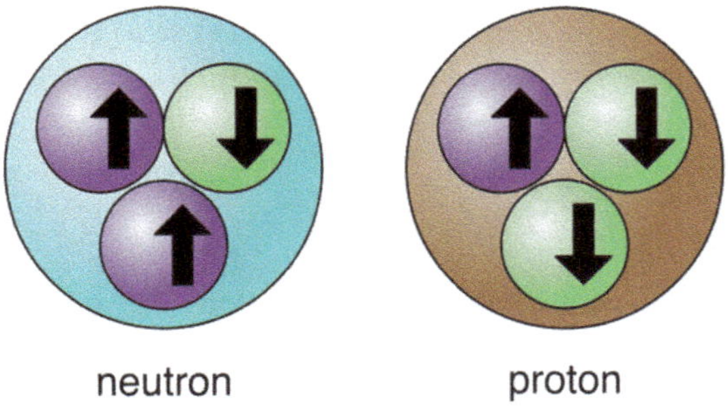

Figure 0.3. Comparison of the quark states in neutrons and protons.

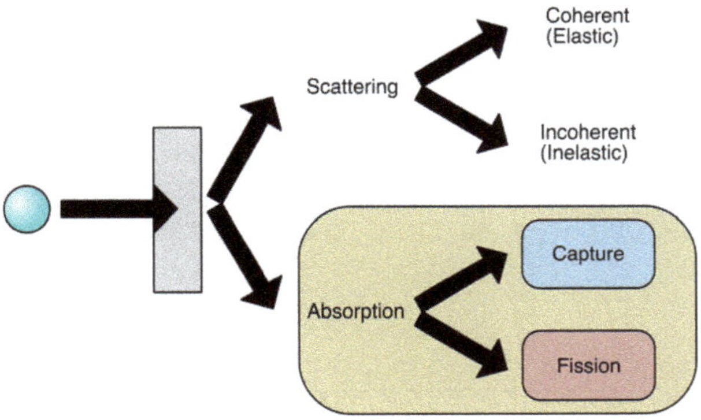

Figure 0.4. The four main types of interaction between neutrons and matter.

generate the heat that is the challenge, and in the case of nuclear power this involves the elusive particle that is the neutron.

The magic of the neutron

The neutron is one of the core components within an atom, the other two are the proton, which is positively charged, and the electron, which is negatively charged. The neutron itself comprises two down quarks (fundamental sub-atomic particles) and one up quark, similar to a proton, which comprises one up and two down quarks, as shown in figure 0.3.

Neutrons have multiple properties, for example they are neutral in charge, but can be influenced by a magnetic field. As a free entity, the neutron has a very short lifetime, ~15 min is the average half-life, whereas the proton is generally considered to be stable (otherwise there would be no hydrogen).

Neutrons can interact with matter in multiple ways, with there being two main types of interaction, scattering and absorption. These two then break down into another two, giving rise to four principal types of interaction—elastic, inelastic, absorption and fission (a schematic is shown in figure 0.4)—and within each of these classifications there are sub-areas. For example, absorption can involve a transmutation from one isotope to another, e.g. ^{135}Xe to ^{136}Xe, or work through what is described as a transfer reaction, where one neutron is absorbed, giving rise to the release of two neutrons, which is described as an (n,2n) reaction, e.g. ^{206}Pb to ^{205}Pb.

$$^{135}_{54}Xe + ^{1}_{0}n \rightarrow ^{136}_{54}Xe. \tag{0.1}$$

$$^{206}_{82}Pb + ^{1}_{0}n \rightarrow ^{205}_{82}Pb + 2^{1}_{0}n. \tag{0.2}$$

Whilst there are four main types of interaction, the two of most interest are capture and fission, which are key to the choice of materials. The scattering interactions are important, but for the purpose of energy generation, the transport of neutrons is key and, as will be explained through the book, it is one the key criteria for a nuclear reactor.

As can be seen in figure 0.5, when neutrons pass through matter, if the absorption of the material is too high neutrons will not transfer through, and get to where they can induced fission, or in the case of fusion transmutation.

While we will discuss how neutron absorption affects the materials chosen within a reactor in more depth in later chapters, we can use a simple example to highlight the impact absorption has on the thickness of materials. How thick would a sheet of iron (Fe) need to be to have the same neutron absorption as zirconium (Zr)?

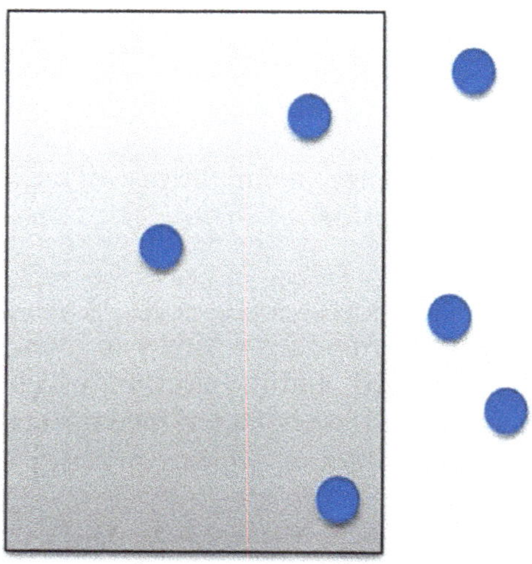

Figure 0.5. Animation of grain growth in impurity free copper plycrystal.

Table 0.1. Absorption data for natural Zr and Fe [2].

Element	Neutron absorption (barns)
Zr	0.185
Fe	2.56

This can be estimated simply by taking a ratio of the absorption of Zr to Fe (table 0.1), so in this case the new thickness would be:

$$\frac{\text{Zr}}{\text{Fe}} = \frac{0.185}{2.56} = 0.072 \times 100 = 7.2\ \mu\text{m}. \tag{0.3}$$

So, as can be seen, having low neutron absorption plays a significant role in the materials chosen within a nuclear reactor, which has led to innovations in the past and means challenges for the future.

This book examines the key challenges that have arisen, discusses the basis for their impact and how the problems can be overcome, and highlights future opportunities. However, if one were to describe the main challenge in nuclear materials as succinctly as possible, the key phrase would be: *neutrons are precious, do not waste them.*

References

[1] IAEA Power Reactor Information System

[2] Chadwick M B *et al* 2006 ENDF/B-VII.0: next generation evaluated nuclear data library for nuclear science and technology *Nuclear Data Sheets* **107** 2931–3060
Chadwick M B *et al* 2011 ENDF/B-VII.1 nuclear data for science and technology: cross sections, covariances, fission product yields and decay data *Nuclear Data Sheets* **112** 2887–996

Acknowledgement

I would like to thank everybody who helped and assisted during the writing, who answered my questions, who read through chapters and who listened to me going through ideas, your help was very much appreciated and welcome. In particular thanks goes to the people at IOP, who helped greatly during the writing, and provided advice when I came across a problem I could not easily solve.

Author biography

Karl Whittle

Karl received obtained his undergraduate degree at the University of Kent, a masters from the University of Aberdeen, and PhD from the Open University. After completing his PhD he undertook postdoctoral appointments at the Universities of Bristol, Cambridge and Sheffield, researching into amorphous materials, and nuclear waste options. He then moved to the Australia Nuclear Science and Technology Organisation (ANSTO), where he led research into the effects on materials of radiation damage. In 2012 he moved back the UK as Senior Lecturer in Nuclear Materials at the University of Sheffield, and in 2015 he moved to the University of Liverpool as the Chair in Nuclear Engineering. Over the years he has developed research linkages across the world, with active collaborations in the US, India, Australia and South Korea.

IOP Publishing

Nuclear Materials Science

Karl Whittle

Chapter 1

Atomic considerations

Before discussing nuclear fission or fusion, it is important to understand the atomic nucleus. An atom comprises positively charged protons and neutral neutrons within the nucleus, and a negatively charged electron cloud surrounding the nucleus balancing the charge, i.e. in a neutral state the proton and electron count are identical. Protons and neutrons are baryons and thus composed of quarks, but the arrangement is different. The proton has two up quarks and one down quark, whereas the neutron has one up and two down, with each species held together by the strong nuclear force.

The atomic nucleus is a balance between the repulsive nature of the protons, with an interaction length of 10^{-2} m, and the attractive nature of the strong force, with an interaction length of 10^{-15} m. Therefore, for the nucleus to remain stable, as the proton number (atomic number) increases, an increasing number of neutrons is required to overcome the repulsion. It is this balancing act that is key to nuclear fission and, to a lesser degree, fusion.

1.1 Isotopes

As outlined above, the number of protons indicates the element, for example hydrogen (H) has one proton, helium (He) has two and uranium (U) has 92, however, each of these elements can have a different number of neutrons. Using hydrogen as an example, there are three common isotopes (table 1.1).

In general, increasing the atomic number increases the number of potential isotopes, such that for uranium there are ~25, none of which are stable, and some of which have very short lives, while some have much longer lives [2]. Since the chemical nature is determined by the proton number, a shorthand is used when describing elements, based on the atomic symbol (e.g. U for uranium), atomic number (number of protons) and mass (sum of protons and neutrons) of the isotope. Uranium has 92 protons and 143 neutrons, giving a total nucleon count of 235:

$$ {}^{A}_{Z}\mathrm{X} = {}^{235}_{92}\mathrm{U}, \tag{1.1} $$

where A = isotope, Z = proton number and X = element symbol.

doi:10.1088/978-0-7503-1104-5ch1

Table 1.1. Isotopes of hydrogen [1].

Protons	Neutrons	Name	Stable?
1	1	Hydrogen	Yes
1	2	Deuterium	Yes
1	3	Tritium	No (half-life 12.31 yr)

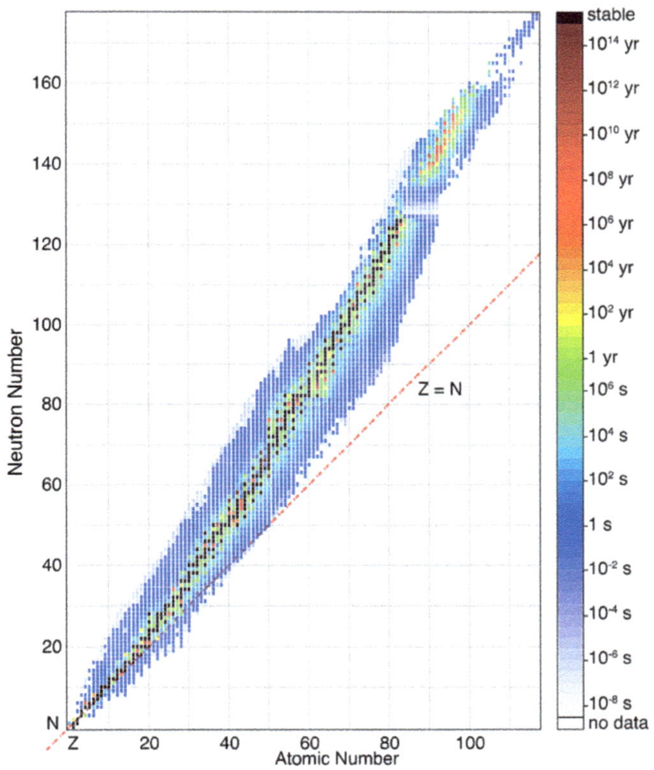

Figure 1.1. Diagram showing the stability of isotopes with increasing atomic number. Data taken from [2].

1.2 Nuclear stability and radioactive decay

As the number of protons and neutrons increases, an indication of likely stability is the ratio of neutron to proton, and this can be split into three main regions, graphically shown in figure 1.1:

(i) at low Z the ratio is close to 1;
(ii) with increasing Z the ratio increases to ~1.5;
(iii) with a Z higher than 83, bismuth is the last element with a stable isotope, there are no naturally stable isotopes after this (all are radioactive, but some have long-lived isotopes), i.e. polonium and above are all radioactive.

The key mechanism by which isotopes can increase their stability is to undergo radioactive decay until a stable isotope is reached. There are five main processes involved in decay, three of which give rise to a change in atomic number (transmutation). These are outlined below:

(i) alpha decay (α-decay)—the ejection of an alpha particle, a ^{4}He nucleus, from the nucleus of the decaying atom;
(ii) beta decay (β-decay)—the ejection of an electron from the nucleus, arising from the internal conversion of a neutron to a proton;
(iii) beta^{+} decay/positron emission—the inverse of beta decay, it results from the conversion of a proton to a neutron, giving rise to the release of an anti-proton, or positron;
(iv) electron capture—a process similar to positron emission, but where an electron is captured from an inner orbital and converts a proton to a neutron;
(v) gamma emission (γ-ray)—emission of a quantised electromagnetic wave from the nucleus of the atom, arising from internal energy transfer within the nucleus;
(vi) spontaneous fission—where the nucleus spontaneously splits into two or more particles.

All of the above mechanisms have relevance when discussing nuclear fission and fusion, however, α-decay and β-decay are more relevant for nuclear materials science, as these processes can give rise to interesting material challenges.

1.3 Alpha-decay (α-decay)

As outlined above, α-decay is the ejection of an α-particle (^{4}He nuclei) from the nucleus, changing both the atomic mass and number. It is a process generally found in the higher elements, such as ^{239}Pu and ^{235}U. After α-decay, the source atom mass decreases by four, and the atomic number by two:

$$^{239}_{94}Pu \rightarrow ^{235}_{92}U + ^{4}_{2}He + \gamma. \tag{1.2}$$

Alpha-decay is relevant in many areas within nuclear materials, such as the formation of He bubbles within nuclear fuel and induced radiation damage in nuclear waste.

1.4 Beta-decay (β-decay)

Beta-decay is the emission of an electron from within the nucleus, through conversion of a neutron to a proton. Using the common fission product of ^{137}Cs as an example, upon beta-decay ^{137}Ba is formed, an increase in atomic number but not in atomic mass:

$$^{137}_{55}Cs \rightarrow ^{137}_{56}Ba + e^{-} + \gamma. \tag{1.3}$$

It can happen across the entire periodic table (e.g. the decay of ^{3}H to ^{3}He) and is useful as a source of fuel for fusion cores or in the conversion of ^{238}U to ^{239}Pu within fission cores. As with α-decay, β-decay can be problematic due to transmutation, which in some cases, such as nuclear waste immobilization, can be highly problematic.

1.5 Beta$^+$/positron emission or electron capture

These processes are similar, involving the conversion of a proton to a neutron, and while the mechanism of the conversion may differ, the end results are the same, i.e. the conversion from element to another.

$$^{67}_{31}Ga + e^- \rightarrow ^{67}_{30}Zn + \gamma. \quad (1.4)$$

$$^{22}_{11}Na \rightarrow ^{22}_{10}Ne + e^+ + \gamma. \quad (1.5)$$

These processes have multiple uses, not least in medical imaging.

1.6 Gamma emission

Gamma emission is the emission of quantised electromagnetic radiation from the nucleus of the atom. It can have a wide range of energies, however, they are typically greater than ~100 keV. The emissions arise from the relaxation of energy states in the nucleus; using as an example the beta decay of ^{99}Mo, the mechanism shown in figure 1.2 occurs.

In this pathway, ^{99}Mo decays via beta mechanisms, with a half-life of 66 h, from ^{99}Mo direct to ^{99}Tc with an efficiency of ~12.5%, and from ^{99}Mo to a metastable form of ^{99}Tc, ^{99m}Tc with an efficiency of ~87.5%. This form of ^{99}Tc has a higher internal energy and decays to the lower energy state through the emission of a gamma ray, with a half-life of ~6 h. Technetium then undergoes a further beta decay, giving rise to ^{99}Ru, a stable isotope.

This mechanism is used widely in medicine for radioactive tracing.

1.7 How do the mechanisms relate to each other?

Each of the mechanisms above, and shown schematically in figure 1.3, with the exception of gamma emission, give rise to a change of atomic number, thus the element changes.

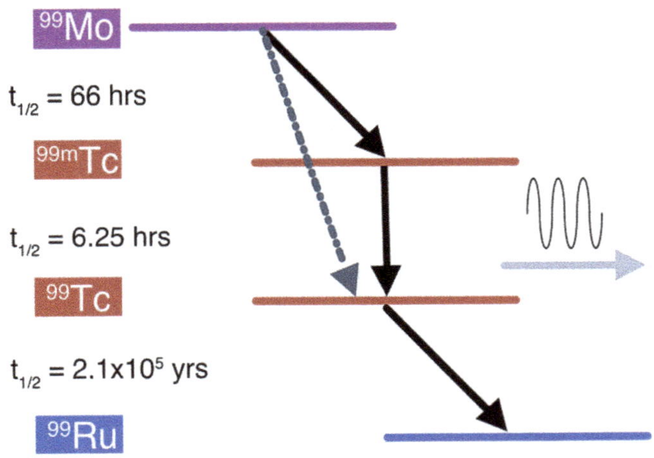

Figure 1.2. Schematic of the decay scheme giving rise to the formation of ^{99m}Tc/^{99}Tc from ^{99}Mo.

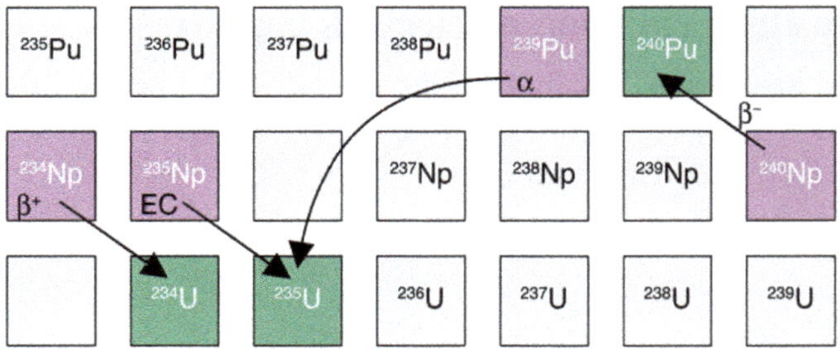

Figure 1.3. Examples of radioactive decays.

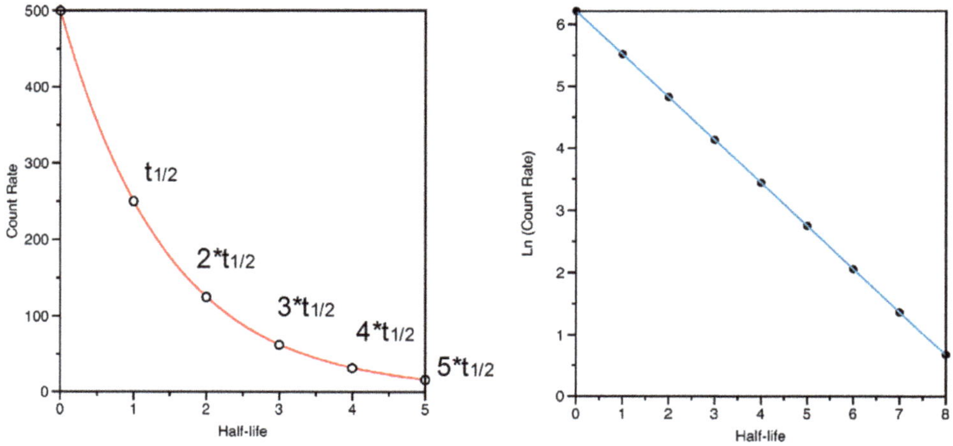

Figure 1.4. (a) Exponential decay of radioactivity predicted by (1.6) and (b) linear logarithmic plot of (1.6).

1.8 Radioactive half-life

The radioactive decay process is probabilistic in nature and occurs at a constant rate, i.e. decays per unit of time. If this is coupled with the amount of material, the following can be derived, based upon the initial amount of isotope present:

$$A(t) = A_0 e^{-\lambda t} \Rightarrow \ln(A) = \ln(A_0) - \lambda t, \tag{1.6}$$

where A_0 is the initial activity, λ is the decay constant and t is time. How do we obtain the decay constant and therefore the half-life?

If activity within a sample is plotted as a function of time, a decaying exponential is found, i.e. e^{-x}, as shown in figure 1.4(a), and thus as a result if natural logarithms are taken a linear line is found, as in figure 1.4(b). The gradient of this decay process is equal to the decay constant for the isotope, which is directly proportional to the radioactive half-life, through the relationship:

$$\lambda = \frac{\ln(2)}{t_{1/2}}, \tag{1.7}$$

where $t_{1/2}$ is the radioactive half-life for the isotope.

Not every half-life is the same, in fact very few if any are identical to one another, they range from 10^{-24} s to 10^{24} years. This leads to the observation that some elements, while appearing naturally stable, are radioactive but have very long half-lives, e.g. ^{232}Th, which is 100% naturally occurring but has a half-life of 14×10^9 years.

1.9 Decay series

Just as an element can undergo decay, it is often the case that the daughter isotope, for example ^{235}U from ^{239}Pu, can also undergo decay, in this case to form ^{231}Th from ^{235}U. Such processes lead to decay pathway, or series, for radioactive isotopes to reach stability. In the actinides, which are common in nuclear fission, there are three common decay series:

(i) uranium—starting with ^{238}U;
(ii) actinium/plutonium—starting with ^{235}U, a decay product of ^{239}Pu;
(iii) thorium—starting with ^{232}Th, the naturally occurring isotope.

Using the actinium series as an example (figure 1.5), the series is predictable, with a family of decays, both α and β in nature, but only in a probabilistic sense. This probability arises from the nature of radioactive decay, which can often be a

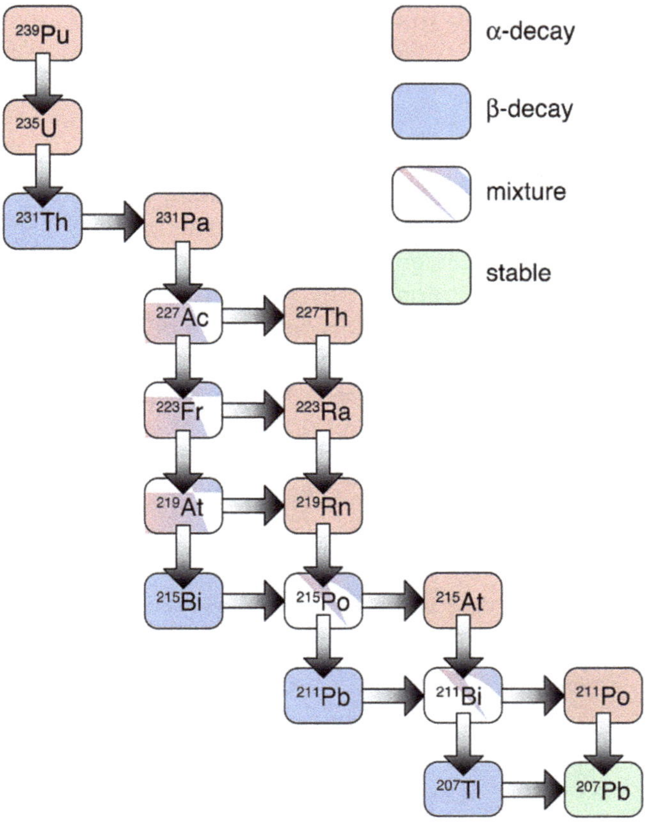

Figure 1.5. Actinium decay series with ^{239}Pu as the starting point.

competing process between two different decay modes. For example, as outlined above in the decay of ^{99}Mo (figure 1.2), there are two routes to ^{99}Ru via ^{99}Tc. One route gives rise to a metastable isotope (^{99m}Tc) with a half-life of 6 h, which then decays to ^{99}Tc. The other method is direct decay to ^{99}Tc, which then has a half-life of 2.1×10^5 years, forming ^{99}Ru. This is just one example of competing decay routes, with in this case the same end product, however, what would happen if the two methods for decay were α and β?

When this competing process is found, again probability helps in describing the nature of the processes and what the decay products will be.

Again using the actinium series as an example, as decay progresses, the pathway is predictable, however, at ^{227}Ac two methods for decay are found, 98.62% β, and 1.38% α. This difference may not seem a significant deviation away from 100% β, and in this case the assumption could be made for 100% β, but in other cases it is more significant. For example, ^{229}Np has a split of 51% α and 49% β+. It can become a significant hindrance during the operation of a fission reactor if the daughter isotopes have differing neutron absorptions or fission yields.

1.10 Observations on isotope stability

It is found that ~60% of all stable isotopes have an even number of protons and neutrons, and for elements of atomic number 20 or greater, they contain more neutrons than protons. The breakdown of stable isotopes and even–odd proton–neutron counts can be seen in table 1.2.

1.11 Binding energy

If radioactive decay enhances the stability of a nucleus, what energy is released and is it calculable? Obviously the energy can be calculated, but with the proviso that it is just an estimate, and for exact determination the process would need to be measured.

The standard mass for a proton is 1.672622×10^{-27} kg and for a neutron it is 1.674927×10^{-27} kg, but to make life easier we shall work in atomic mass units (amu), which have the mass 1 amu = 1.660539×10^{-27} kg. Thus the mass of a proton = 1.007276 amu and the mass of a neutron = 1.008665 amu.

Using ^{4}He as an example, what would the expected mass be? Since ^{4}He contains two neutrons and two protons, the mass would be expected to be 4.031882 amu, however, the mass of ^{4}He is found to be 4.002603 amu, a difference of 0.029279 amu.

Table 1.2. Number of odd/even isotopes and examples for each. Data taken from the table of isotopes [1].

Protons	Neutrons	Stable no	Example
Even	Even	159	^{4}He, ^{16}O
Even	Odd	53	^{90}Zr, ^{99}Ru
Odd	Even	50	^{133}Cs, ^{7}Li
Odd	Odd	4	^{6}Li, ^{39}K

This difference is significant, as by applying the conservation of mass principle, there should be no difference. The difference is due to the mass being converted to energy and used to help bind the atom, hence the term 'binding energy'. It is possible to calculate the binding energy in the system by using the difference in masses and applying Einstein's relationship, $E = mc^2$. Rather than working in base SI units of kg, we will work in amu units and use the relationship that 1 amu = 931.494 MeV c^{-2}.

The mass of ^{4}He = 4.002602 amu.

The mass of two protons + two neutrons = 4.031882 amu.

The mass difference = 0.029279 amu.

This leads to the binding energy within the nucleus being:

$$E = mc^2 = 0.029279 \times 931.494$$
$$E = 27.27 \text{ MeV} \tag{1.8}$$

Therefore in a ^{4}He nucleus (α-particle) the binding energy, i.e. the energy contained within the nucleus keeping it together, is 27.27 MeV, equating to ~ 6.8 MeV per nucleon within the nucleus (proton or neutron). If this approach is taken with all the nuclei in the periodic table, a trend arises that has importance for nuclear fission and fusion (figure 1.6).

As can be see in figure 1.6, the binding energy per nucleon is highlighted in three regions. The first region shows a general increase in binding energy per nucleon, reaching a maximum at Fe/Ni, which itself has implications for nuclear fusion. It then decreases slowly as the atomic number increases, until above Po Coulombic repulsion becomes so great that fission becomes easier.

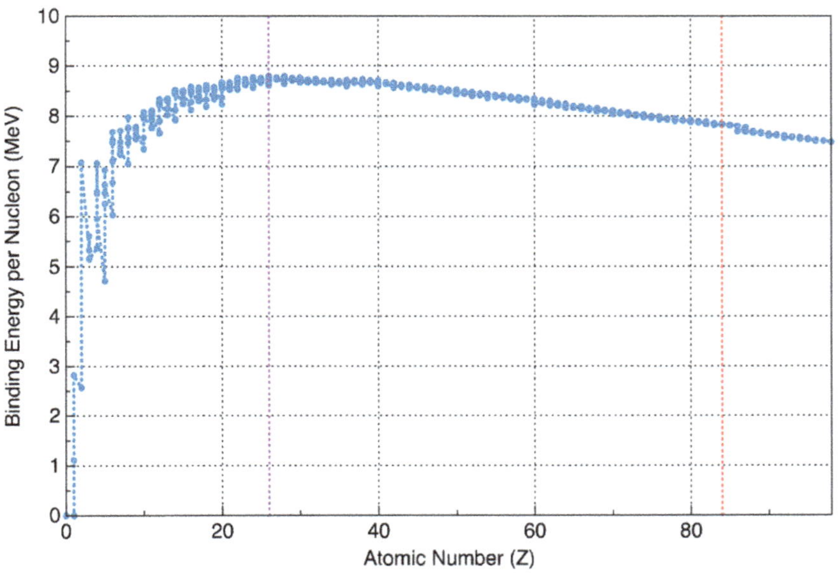

Figure 1.6. Diagram showing the relative stability per nucleon for the elements where data have been determined. Data taken from [2].

1.12 Fission and fusion

Before discussing the processes for nuclear fission and fusion it is important to define them:

- fission—the process by which an object splits into two or more parts, shown in the animated version of figure 1.7;
- fusion—the joining of two or more objects, forming one, shown in the animated version of figure 1.8.

These processes are the inverse of each other in terms of nuclear fission and nuclear fusion and the definitions can be given as:

- nuclear fission—the splitting the of an atomic nucleus into two or more lighter nuclei;
- nuclear fusion—the joining of two atomic nuclei into a heavier nucleus.

Schematics for the processes are given in figures 1.7 and 1.8.

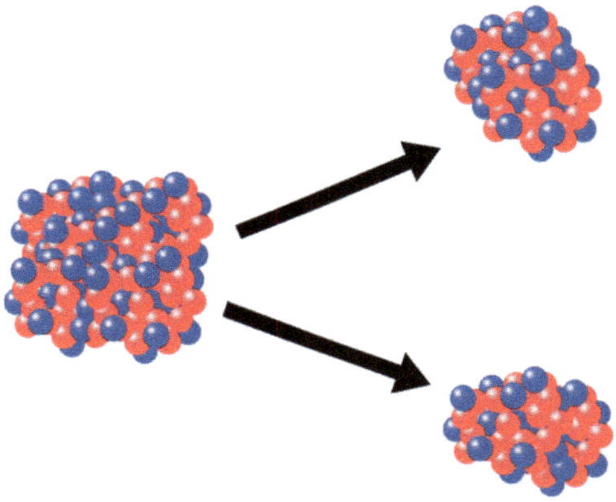

Figure 1.7. Schematic and animation of the fission process.

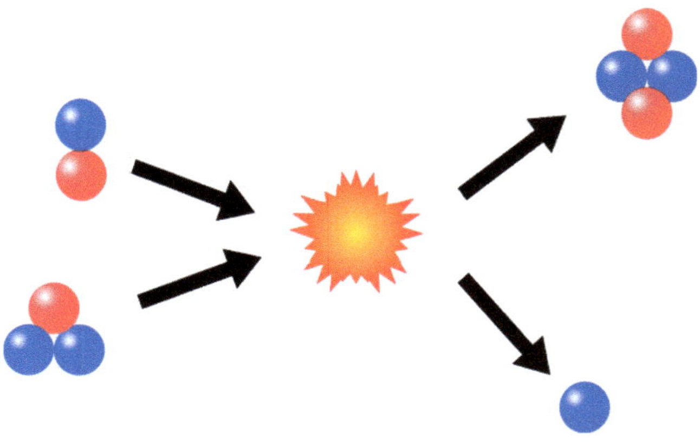

Figure 1.8. Schematic and animation of the fusion process.

1.13 Spontaneous fission

In the discussion of radioactive decay one mechanism was mentioned but not discussed, spontaneous fission. This is nuclear fission in its simplest form, it arises from a non-zero probability of fission with respect to the strong force holding the nucleus together. For example, in the spontaneous fission of ^{238}U to lanthanum and bromine, with the release of three neutrons, what is the energy released by this process? Using the same method outlined previously for α-decay:

$$^{238}_{92}U \rightarrow {}^{142}_{57}La + {}^{90}_{35}Br + 3\ {}^{1}_{0}n \tag{1.9}$$

mass of ^{238}U = 238.029181 amu
mass of ^{145}La = 144.92163 amu
mass of ^{90}Br = 89.93063 amu
mass of neutron = 1.00866 amu

$$\begin{aligned} \text{mass change} &= {}^{238}U - {}^{145}La - {}^{90}Br - [3 \times {}^{1}n] \\ &= 238.029181 - 144.92163 - 89.93063 - [3 \times 1.00866] \\ &= 0.1516\ \text{amu} \end{aligned}$$

$\therefore$ energy released = 0.1516 × 931.494 = 141.3 MeV.

The energy released by this spontaneous fission process is 141.3 MeV, a not insubstantial amount of energy. So is this the process by which nuclear fission occurs in a reactor core? Unfortunately not, while spontaneous fission can occur, and indeed does occur, the probability of it happening is not sufficiently high to be used to generate electricity without further assistance, examples are shown in table 1.3.

Since the probability of spontaneous fission is very low, can we do anything to increase the likelihood of fission? Obviously, the answer is yes, otherwise most nuclear fission-based reactors would close pretty quickly. The key to increasing the likelihood of fission can be found in the process shown above, the release of neutrons.

Table 1.3. Spontaneous fission probabilities for selected elements [1, 2].

Isotope	Half-life (yr)	Fission probability per decay	Neutron yield
^{235}U	7.04×10^{8}	7×10^{-11}	1.86
^{238}U	4.47×10^{9}	5.4×10^{-7}	2.07
^{239}Pu	2.41×10^{4}	3.1×10^{-12}	2.16
^{240}Pu	6569	5.7×10^{-8}	2.21
^{250}Cf	2.638	0.74	3.73

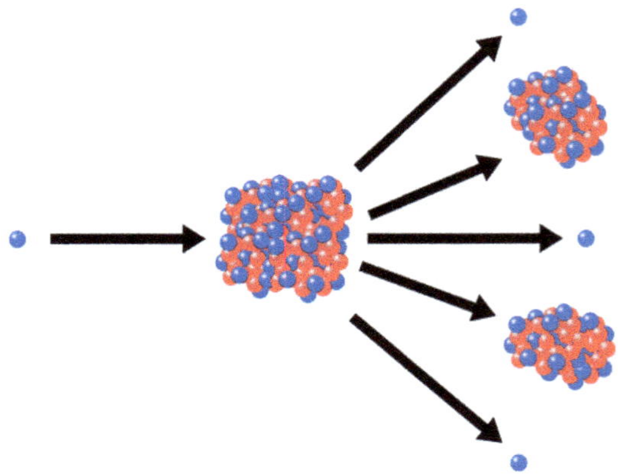

Figure 1.9. Schematic and animation of induced nuclear fission.

1.14 Inducing fission and chain reactions

For some nuclei it is possible to induce instability through the capture of a neutron, which leads to the nucleus splitting to reduce the instability. For example, inducing fission in ^{235}U with a neutron can lead to the process below, with an expected energy release of ~173 MeV.

$$^{235}_{92}\mathrm{U} \rightarrow {}^{137}_{55}\mathrm{Cs} + {}^{96}_{37}\mathrm{Rb} + 3^{1}_{0}\mathrm{n}. \tag{1.10}$$

The three neutrons released by this process can then induce further nuclear fission in atoms nearby, leading to a chain reaction of fission. This process of inducing fission, releasing neutrons to induce further fission, is the principle by which nuclear fission reactors work. See figure 1.9 for a schematic of induced nuclear fission.

1.15 Neutron absorption and fissile and fertile isotopes

Not every isotope will absorb a neutron and undergo fission, in many cases the neutron will be absorbed and not undergo fission, for example ^{155}Gd (one of the largest neutron absorbers) absorbs a neutron, forming stable ^{156}Gd. Such an absorption can be used advantageously, for example in moderating the reactor.

$$^{155}_{64}\mathrm{Gd} + {}^{1}_{0}\mathrm{n} \rightarrow {}^{156}_{64}\mathrm{Gd}. \tag{1.11}$$

A fissile isotope is one that can undergo nuclear fission, e.g. ^{235}U and ^{239}Pu, which form the basis of most currently used nuclear fuel. Such isotopes tend to have an odd number of neutrons, see table 1.2 and the example below.

$$^{235}_{92}\mathrm{U} + {}^{1}_{0}\mathrm{n} \rightarrow {}^{137}_{55}\mathrm{Cs} + {}^{96}_{37}\mathrm{Rb} + 3^{1}_{0}\mathrm{n}. \tag{1.12}$$

A fertile isotope is one that does not undergo fission, but which transforms into one that does; these tend to have an even number of neutrons (see table 1.2). They are used in nuclear fuel and provide a means by which fuel can be generated within a reactor core. For example, ^{238}U can be converted to ^{239}Pu, and ^{232}Th to ^{233}U—both processes going from non-fissile to fissile.

$$^{238}_{92}\text{U} + ^{1}_{0}\text{n} \longrightarrow ^{239}_{92}\text{U} \xrightarrow[23.5\text{ min}]{\beta} ^{239}_{93}\text{Np} \xrightarrow[2.36\text{days}]{\beta} ^{239}_{94}\text{Pu}. \tag{1.13}$$

A similar process occurs for ^{232}Th, leading to the formation of ^{233}U, which is then fissile and can undergo fission.

1.16 Increasing fission yield

As has been outlined, spontaneous fission generally has a low probability, particularly in isotopes that are found naturally. One way to increase fission is to introduce a source of neutrons from an external source, which, while possible, is not ideal. There are, however, two other methods that can greatly enhance the probability of fission and allow the chain reaction to continue.

The easiest method for increasing the likelihood of fission is to increase the relative content of fissile isotopes in the fuel, i.e. enrichment. Using uranium as an example, natural uranium contains ^{238}U, ^{235}U and ^{234}U at differing levels (table 1.4), and these nuclei have different fissile/fertile behaviours. Since ^{235}U is the only naturally occurring fissile nucleus, the other two being fertile, the likelihood of fission is small. However, it can and does occur under the right conditions, for example in the CANDU and MAGNOX reactors, if the reactor core is designed appropriately. What is easier is to increase the relative amount of ^{235}U with respect to ^{238}U, i.e. enrichment. As the level of ^{235}U increases, the ability of the core to sustain a nuclear fission process increases, and can become self-sustaining, once a critical mass of fissile material is reached. There are multiple approaches to enrichment, each of which has advantages and disadvantages, but they are out of context here. It is sufficient to say that enrichment is a complex process fraught with many challenges; separating ^{238}U and ^{235}U, i.e. a mass difference of three in 235 (~1.2%), is not easy.

A fissile isotope on absorption of a neutron does not always fission, it is only one of the possibilities available, for example a further neutron capture could occur, moving the isotope from fissile to fertile or even stable, i.e. transmutation. The degree of fission/transmutation depends on multiple factors, but the most dominant is that of incident neutron energy: if the energy is high, transmutation is more likely, if it is low, fission is more likely. Therefore, to enhance the likelihood of fission, neutrons can be moderated from high energy, allowing them to be captured and induce further fission. The fission cross-section (a measure of fission probability) is shown in figure 1.10 for ^{235}U in the energy range of 1 keV–1 MeV.

Table 1.4. Natural abundance of uranium isotopes, their fission/fertile nature and what they become if fertile [1].

Isotope	Half-life (yr)	Relative amount (%)	Fissile/fertile
^{234}U	2.46×10^5	0.0054	Fertile (^{235}U)
^{235}U	7.04×10^8	0.7204	Fissile
^{238}U	4.47×10^9	99.2742	Fertile (^{239}Pu)

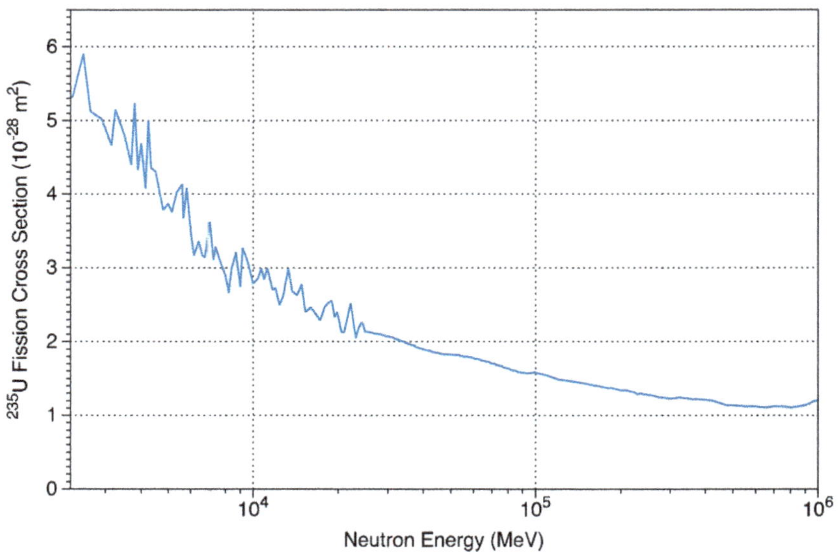

Figure 1.10. Fission cross-section for ^{235}U with varying energy. Data taken from [2].

As can be seen in figure 1.10, the general trend is for less fission with increasing energy, and this can have multiple uses, not only for increasing fission yield through moderation, but also for transmuting one isotope to another, i.e. breeding isotopes. One common method for moderating (increasing the likelihood of fission) is to use a material/element that does not absorb neutrons easily, but can absorb some of the energy. Two common methods use H_2O or C, often in the form of CO_2, and these have a low neutron abscrption and thus do not readily absorb neutrons, but do slow them down.

The energy arising from fission is ultimately converted from kinetic energy, i.e. particles moving with energy, into heat. As a consequence, H_2O- (more commonly) and CO_2-cooled reactors have been developed that take advantage of the ability of both to transfer heat away from the core, while at the same time moderating neutrons. These thermofluids then transfer the heat from the core and use it drive turbines, generating electricity.

1.17 What are the key criteria for nuclear fission?

It is can often be the case that nuclear fission is considered to be a technically difficult process to undertake, whereas in reality it is not as difficult as expected. The key factor in keeping a nuclear reactor running is keeping the fission chain reaction proceeding; this is the technical challenge. The key criteria for running a nuclear reactor are broken down into two groups, the required and the desirable.

1.17.1 Required components

- Fuel—this can either be fissile material, e.g. ^{235}U and ^{239}Pu, or fertile material, e.g. ^{238}U and ^{232}Th.

- Low neutron absorption—for a chain reaction to proceed, it is obvious that the neutrons must not be impeded or absorbed, where possible. As such, there are many isotopes that would, in an ideal world, never be used within a core, e.g. ^{6}Li, ^{10}B, ^{113}Cd, ^{174}Hf, ^{155}Gd and ^{157}Gd (highest naturally occurring neutron absorption).

1.17.2 Desirable components

- Moderation of neutrons—increasing the fission yield upon neutron absorption, it can also be coupled with the removal of the heat from the core.
- Enrichment—as with moderation, this increases the efficiency of the fission process by increasing the concentration of fissile material in the core.

Thus, there are routinely four variables in a fission core that need to be managed, each with their own factors that can change during reactor operation. The key point to remember is that each of the criteria have material challenges that need to be overcome, and these are the focus for the remainder of this text.

References

[1] Wieser M E *et al* 2013 Atomic weights of the elements 2011 (IUPAC Technical Report) *Pure Appl. Chem.* **85** 1047–78
[2] Chadwick M B *et al* 2006 ENDF/B-VII.0: next generation evaluated nuclear data library for nuclear science and technology *Nuclear Data Sheets* **107** 2931–3060
[3] Chadwick M B *et al* 2011 ENDF/B-VII.1 nuclear data for science and technology: cross sections, covariances, fission product yields and decay data *Nuclear Data Sheets* **112** 2887–996

IOP Publishing

Nuclear Materials Science

Karl Whittle

Chapter 2

Radiation damage

During fission or fusion, one key outcome, other than the desired release of energy, is the release/formation of energetic particles. In fission they can have a range of masses and energies, dependent upon the isotope being split. In fusion the situation is much simpler because there are only a few particles that can be released, as the starting isotopes have much less mass to form from.

$$^{238}_{92}\mathrm{U} \rightarrow {}^{145}_{57}\mathrm{La} + {}^{90}_{35}\mathrm{Br} + 3^{1}_{0}\mathrm{n}. \tag{2.1}$$

For example, in the fission process outlined in chapter 1, and shown in figure 2.1, for the spontaneous fission of ^{238}U, what was originally one atom has now split into two smaller atomic nuclei (^{145}La and ^{90}Br) and three neutrons. Each of these released particles has kinetic energy, a total of ~141 MeV, and therefore momentum, and it is how each of these particles interacts with the surrounding atoms that influences the long-term stability of the material.

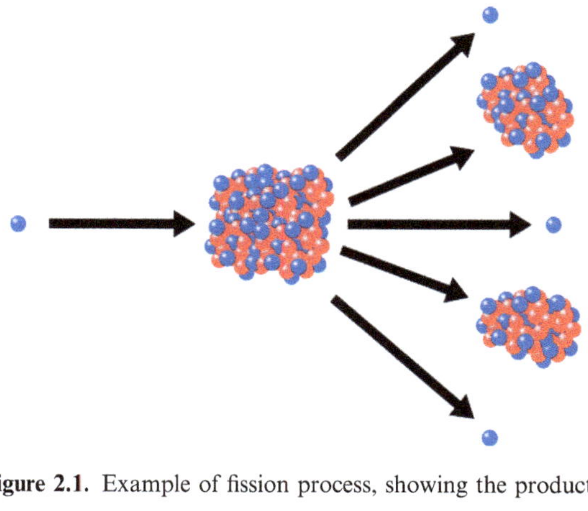

Figure 2.1. Example of fission process, showing the products.

doi:10.1088/978-0-7503-1104-5ch2

2.1 Key definitions

In order to understand what is involved in radiation damage, we need to define some key parameters that will be used during the discussion.

2.1.1 Primary knock-on atom (pka)

This is the incident, or primary, damaging particle, which can be an atom or neutron, which has a corresponding energy (eV).

2.1.2 Threshold displacement energy (TDE)

This is the energy required to displace an atom from its lattice position. This can be estimated in one of two ways: using a non-relativistic method, or one that takes into account the effects of relativity. For simplicity, we will consider the non-relativistic method. The kinetic of energy of a particle (KE_{max}) is usually considered, arising from the energy of the incident particle (E_{p}) and the masses of the impacting ion and the lattice ion, respectively (m_{ion} and m_{lat}).

$$KE_{\text{max}} = E_{\text{p}}\left[\frac{4m_{\text{ion}}m_{\text{lat}}}{(m_{\text{ion}} + m_{\text{lat}})^2}\right]. \tag{2.2}$$

For example, how much energy would be required to displace a Fe atom from its atomic position in steel, with a threshold displacement energy of 50 eV, when hit by a He nuclei (alpha particle)? To answer this question we only need to know one piece of information, the threshold displacement energy, which in this case is 50 eV. Since the mass of He is 4, and the mass of Fe is 56, the calculation becomes:

$$E_{\text{p}} = KE_{\text{max}}\left[\frac{(m_{\text{ion}} + m_{\text{lat}})^2}{4m_{\text{ion}}\ m_{\text{lat}}}\right] = 50\left[\frac{(4 + 56)^2}{(4 \times 4 \times 56)}\right]. \tag{2.3}$$
$$E_{\text{p}} = 200\ \text{eV}$$

Therefore, for He to displace an Fe atom with a TDE of eV, the He would ideally need to have 200 eV of energy.

The next definition concerns how much damage has occurred within a system and for this we use the term 'displacements per atom'.

2.1.3 Displacements per atom (dpa)

Put simply, this describes on average how many atoms have been displaced in the lattice, over the whole system. A value of 1 dpa implies that each atom in a system has moved on average once. This can be profound, as it implies that each atom has displaced in the system—consider that for a second. When we come to the impacts of damage in future nuclear materials, where 400 dpa is not unrealistic, this becomes a major materials challenge that needs to be overcome.

For example, in a sample containing 100 Fe atoms, how many atoms have moved if the average effect from He damage is 0.25 dpa? In this case, since the dpa is 0.25, implying that 1/4 of all atoms in Fe sample have been displaced, the answer is 25.

Now imagine this scaled to 4 dpa, each atom will have moved, on average, four times … this is not always easy to predict.

Having defined key parameters in understanding damage, we can now begin to model the damage process itself.

2.2 Radiation damage

Using fission as a source for damage, imagine an incoming particle, for example a He nuclei. This particle has two possibilities when impacting on an atom in the lattice, dependent upon its energy, i.e. pka.

Upon impact, the particle will transfer some/all energy to the target, and if this energy is less than the threshold displacement energy, some transfer of energy will occur, but the pka will not displace the target atom (as shown in figure 2.2).

If the energy of the pka is above the displacement energy, the transfer of energy can result in the target atom being displaced (figure 2.3). Similar to billiards, this transfer of energy may not result in a 100% transfer of energy, the incoming atom have may sufficient energy to continue moving. It is very rare that a 100% energy transfer occurs; using billiards as an analogy, a stun shot is rare.

The process of damage may not be limited to one particle impacting on and displacing another, it can continue for significant lengths of time, like a cannon in billiards. The degree of damage is dependent upon many factors, such as the energy of the pka and the mass of the target. One factor which is common to both is that some of the kinetic energy is mutated into heat, i.e. the local temperature

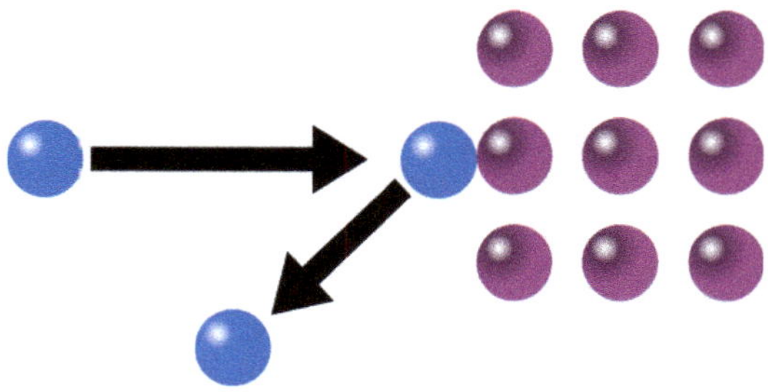

Figure 2.2. Process of atom impacting on lattice without sufficient energy to overcome the threshold displacement energy, see also animation.

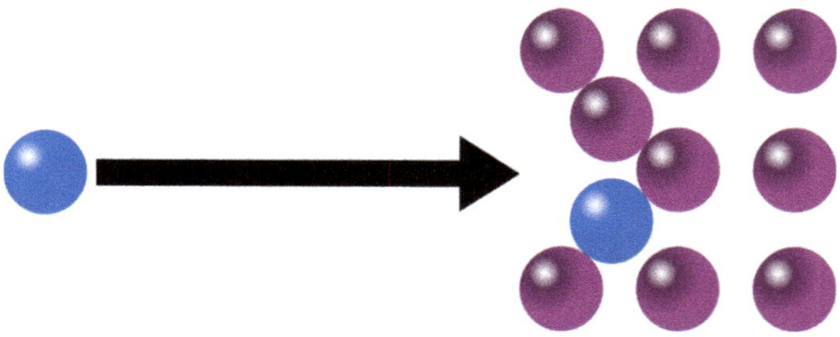

Figure 2.3. Process of atom impacting on lattice with sufficient energy above the threshold displacement energy, see also animation.

around an impact will rise. This will become an issue later, in the discussion on recovery from damage.

Is it possible to predict how much damage will arise from a particle with an energy **X** in a lattice? The answer to this question is yes, with a 'but'. The but comes from the assumptions made and is the basis of the Kinchin–Pease method for predicting damage [1].

2.3 Prediction of damage—the Kinchin–Pease methodology

If the probability of displacement is plotted against the energy of the incident particle using the outline proposed above, we should obtain what is shown in figure 2.4. However, since having a particle with an energy higher than that required to displace an atom does not always result in displacement, what is shown in figure 2.5 is more likely. The differences can be accounted for if one remembers that outside of simulations, lattices are not perfect. For example:

(i) lattice atoms are not fixed, they vibrate on their positions, and such vibrations can modify how they interact with impacting atoms;

(ii) there are likely to be intrinsic defects, such as vacancies/interstitials, in the lattice, along with extrinsic defects, such as impurities;

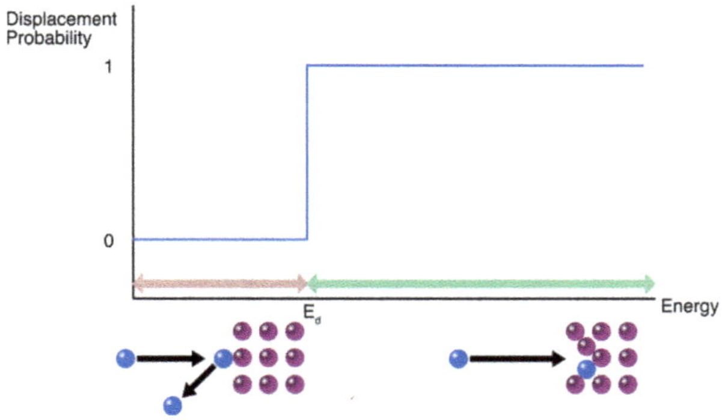

Figure 2.4. Ideal probability for atom displacement with increasing energy of primary knock-on atom, see also animation.

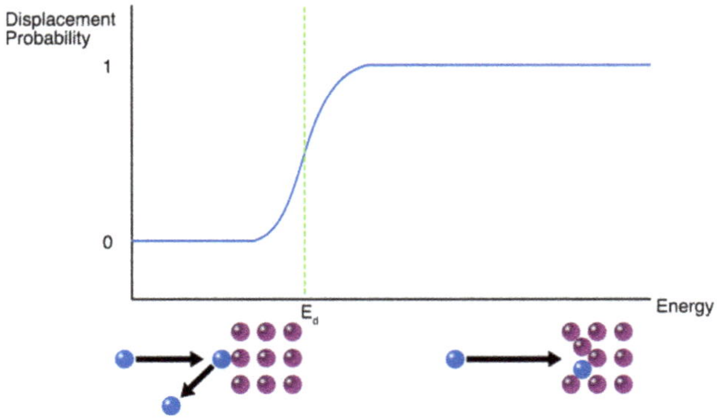

Figure 2.5. More realistic probability for atom displacement with increasing energy of primary knock-on atom, see also animation.

(iii) the orientation of the crystal is a factor—in any structure there are easy and hard directions for atoms to move in.

Each of the above differences is capable of modifying how likely it is that an impacting atom will result in the displacement of a target atom. Therefore, figure 2.5 is a more realistic approach to modelling damage. Is it possible to predict the probability of damage?

2.3.1 Nuclear and electronic stopping factors/powers

There are generally considered to be three interactions that slow, or stop, a high-energy particle in a material: elastic, inelastic and radiation. For the purposes of this discussion, we will ignore radiation and focus on the elastic and inelastic interactions.

Elastic interactions usually occur between nuclei within the structure, i.e. they are collisions between the incoming atom and the target material. Using the example of He into Fe, this would be a He nuclei colliding with an Fe atom. This type of interaction is also known as nuclear stopping and can be compared to playing billiards on a flat smooth surface, i.e. cannon shot.

Inelastic interactions usually occur between the incident particle and the electron cloud within the lattice. Using the example of He into Fe again, this would be the He nuclei interacting with the electron cloud and transferring some of its momentum (energy) to the electronic structure. This type of interaction is also known as electronic stopping and can be compared to walking through water—more difficult than walking on dry ground.

Both interactions eventually stop the incident particle, either by collision or by drag, hence they are known as stopping factors/powers.

2.3.2 How do these powers effect the stopping of the incident particles?

If the electronic/nuclear stopping factors for He into Fe and Fe into Fe are compared (figure 2.6), it can be seen that at low energies, nuclear stopping dominates, whereas at high energies, electronic stopping dominates. The key to being able to predict the likely probability of damage is to understand the change in dominance from nuclear to electronic.

As can be seen in figure 2.6(b), for Fe into Fe, the change from nuclear to electronic stopping occurs at ~700 keV, whereas for He into Fe, the change occurs at ~700 eV. Hence, the majority of the energy loss from He particles is through electronic interactions. These changes directly impact on how a material undergoes damage, and they are a key criterion in the Kinchin–Pease methodology for the prediction of damage.

2.3.3 The Kinchin–Pease methodology for predicting levels of damage

In 1955 Kinchin and Pease [1] proposed a methodology for predicting levels of damage based on three parameters and four assumptions. The three parameters are:

(i) the energy of the primary particle (pka) ($\boldsymbol{T}$);
(ii) the threshold displacement energy for the target material ($\boldsymbol{E_{\mathrm{d}}}$);
(iii) the energy cutoff above which electronic stopping dominates over nuclear stopping ($\boldsymbol{E_c}$).

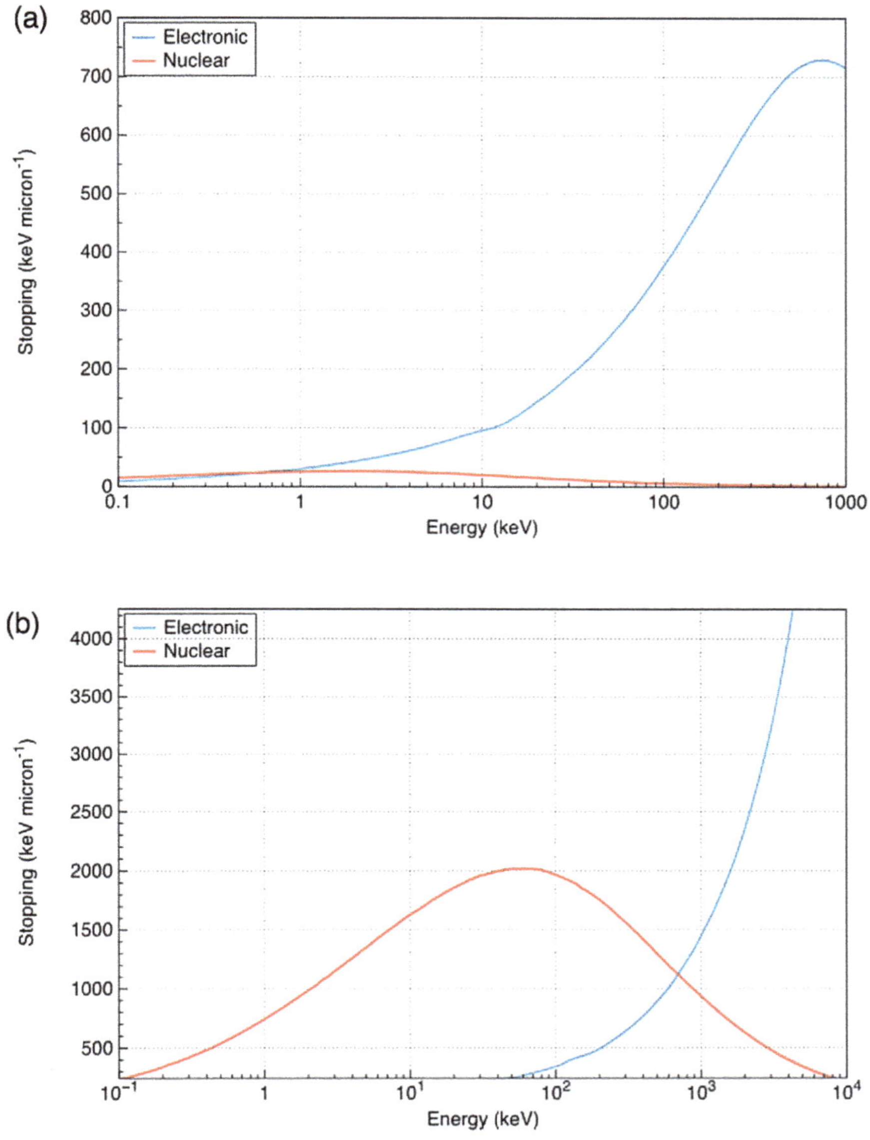

Figure 2.6. The electronic and nuclear stopping powers for (a) He into Fe and (b) Fe into Fe. Powers calculated using SRIM2013 [2].

The key assumptions are:

(i) the damage cascades are elastic and two-body in nature (as in billiards);
(ii) there is no spread in the probability of damage, i.e. $P = 1$ for particles with a kinetic energy $>E_d$;
(iii) there is no energy loss into the lattice, i.e. $T = T' + e$;
(iv) energy loss through electronic loss is described through a simple cut off in energy, $T_{pka} < E_c$;

(v) interactions are described using a hard-sphere (billiard ball) approach;
(vi) the atomic structure is random, i.e. crystal structure is ignored.

How can we use these assumptions to predict the number of atoms displaced? During the first collision in the lattice, energy is transferred (e):

$$N(T) = N(T - \varepsilon) + N(\varepsilon - E_d). \tag{2.4}$$

If we apply assumption (iii), which results in no energy being transferred to the lattice,

$$N(T) = N(T - \varepsilon) + N(\varepsilon). \tag{2.5}$$

How can we estimate e? As both the pka and lattice atoms are considered to be the same in this methodology, it is possible to estimate e, as assuming a uniform distribution of energy transfer, the following relationship can be used,

$$(\varepsilon, \varepsilon - d\varepsilon) = \frac{d\varepsilon}{T}, \tag{2.6}$$

which can now be applied to equation (2.5):

$$\begin{aligned} N(T) &= \int_0^T (N(T - \varepsilon) + N(\varepsilon))\frac{d\varepsilon}{T} \\ N(T) &= \frac{2}{T}\int_0^T N(\varepsilon)\mathrm{d}\varepsilon \end{aligned}. \tag{2.7}$$

This can be solved by splitting into integrals with modified variables, based on assumption (ii), i.e.

$$\begin{aligned} N(T) &= 0, \ \ T < E_d \\ N(T) &= 1, \ \ E_d < T < 2E_d. \end{aligned} \tag{2.8}$$

If these are combined with (2.5) and split into three ranges, we obtain the following:

$$\begin{aligned} N(T) &= \frac{2}{T}\int_0^T N(\varepsilon)\mathrm{d}\varepsilon \\ N(T) &= \frac{2}{T}\left[\int_0^{E_d} 0 \ \mathrm{d}\varepsilon + \int_0^{2E_d} 1\mathrm{d}\varepsilon + \int_{2E_d}^T N(\varepsilon)\mathrm{d}\varepsilon\right] \\ N(T) &= \frac{2E_d}{T} + \frac{2}{T}\int_{2E_d}^T N(\varepsilon)\mathrm{d}\varepsilon. \end{aligned} \tag{2.9}$$

If we multiply by T and differentiate with respect to T, plus a few other steps, we end up with $N(T) = \frac{T}{2E_d}$, in the energy range of $2E_d < T < E_c$, i.e. assumption (iv). The key information to derive from this outline is not the derivation itself, but the outcomes and how they are used.

The implication of the assumptions and parameters are as follows:

(i) below the threshold displacement energy (E_d), the probability of damage is 0, i.e. $N(T) = 0$;

(ii) between the minimum and twice the threshold displacement energy ($2E_d$), the probability is 1, i.e. $N(T) = 1$;

(iii) at energies above $2E_d$ the probability of displacement is $N(T) = \frac{T}{2E_d}$;

(iv) at energies above E_c the probability of displacement remains constant, i.e. $N(T) = \frac{E_c}{2E_d}$, however, once the incident particle has lost energy through electronic interactions, assumption (iii) comes back into play.

Figure 2.7 highlights the outcome of the Kinchin–Pease methodology.

2.3.4 Modifications to Kinchin–Pease

There have been a few modifications of the Kinchin–Pease methodology, generally in the efficiency of displacement. The most commonly used is the

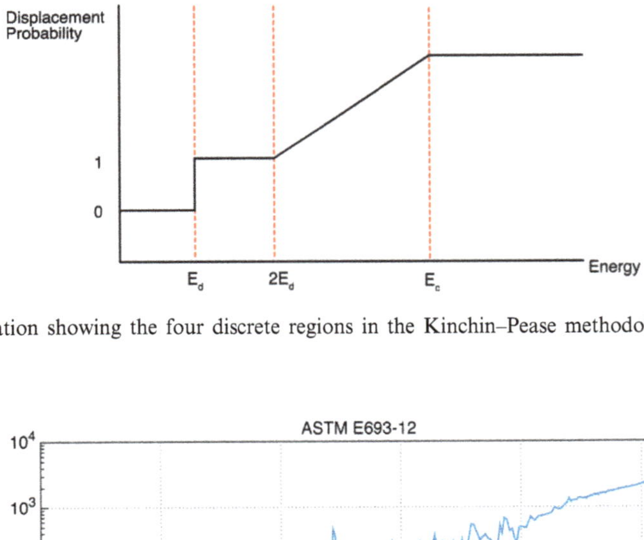

Figure 2.7. Animation showing the four discrete regions in the Kinchin–Pease methodology for predicting radiation damage.

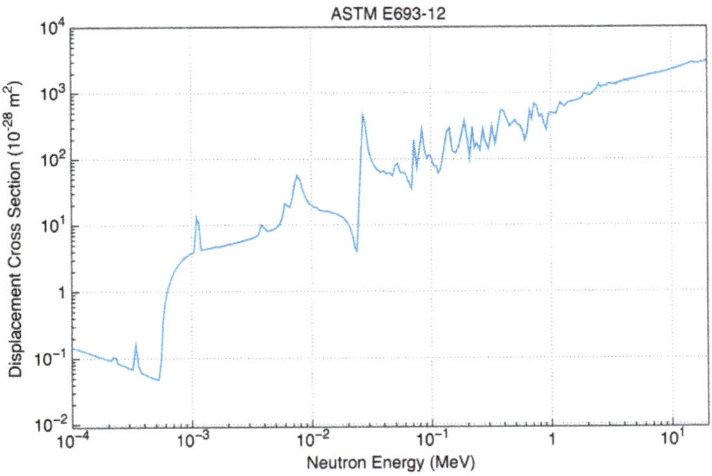

Figure 2.8. Plot showing how the predicted damage cross-section for neutron damage within Fe varies with energy. Original data from [4].

Norgett–Robinson–Torrens (NRT) [3], which modifies it as $N(T) \simeq 0.8(\frac{E_c}{2E_d})$, where $T > 2E_d$, compared with Kinchin–Pease, which is $N(T) = \frac{E_c}{2E_d}$.

NRT is a commonly accepted model for predicting levels of damage in a material and forms the basis of the ASTM standard for predicting irradiation damage in steels [4], and shown graphically in figure 2.8.

It should be noted that both Kinchin–Pease and NRT are ideal cases and assume there is no recovery by the lattice from the damage induced.

2.4 Implications of damage

While it is worthwhile understanding how to model damage in a material, and being able to predict the levels of damage is a key factor, it is not the main concern for materials in a damage-inducing environment. The key concern relates to the end of the damage process and the implications it raises.

If we model the damage track and how it evolves, multiple options are possible, but for the purposes of this discussion we will focus on a simple case of He damaging UO_2, i.e. alpha decay in fuel. Throughout the damage process daughter cascades can form, each of which will have a final state, which will predominantly give rise to interstitial atoms coupled with atomic vacancies in the lattice. How each of these behaves is key to the long-term behaviour of the material. To discuss how these can behave we will look at the three extremes in their behavior: no recovery, full recovery and partial recovery.

2.4.1 No recovery from damage

This case assumes that there are no recovery processes in the system capable of rebuilding the original lattice/structure. The end product of this process is rapid amorphisation, with a concomitant modification in properties, from the desired and predictable to the often broadly unpredictable and highly undesirable.

2.4.2 Full recovery from damage

This is the opposite of no recovery, with the lattice fully recovering from the induced damage, ideally regaining the original lattice/structure.

2.4.3 Partial recovery from damage

This is the hybrid of the two extremes, and the case most often found, where there is partial recovery from damage, but with significant damage remaining.

2.5 Outcomes from damage

2.5.1 Basic recovery

As outlined above, the two key end products of damage are vacancy and interstitial formation, also known as Schottky and Frenkel defects. It is how these behave that is at the core of material behaviour when damaged. Generally with radiation damage, a Frenkel defect will form in conjunction with a Schottky defect, giving rise to a Frenkel pair, shown in figure 2.9.

In a system that recovers from damage, there will probably still be defects present in the lattice, it is how these defects interact that ultimately decides how a material behaves long-term with increasing damage.

For example, on a simplistic level, a vacancy in the lattice can be used to relieve any stress induced through the formation of an interstitial, as can be visualised in figure 2.10. However, this is ideal and does not take into account the other possibilities. What is more likely to occur, is a modified version of this process, which can lead to the formation of a stacking fault in the lattice (figure 2.11). In this case, the lattice has now formed an interstitial defect loop, which can lead to further complications with increasing damage. There is an alternative, inverse, version of this effect, where the lattice combines vacancies in a similar manner: a vacancy defect loop (figure 2.12).

Even taking into account the formation of defect loops, or simple annihilation processes, this is not the end of the problem. When systems such as UO_2 are undergoing fission, or normal alpha-decay process, gas atoms form, whether they be fission gas, i.e. Kr/Xe, or from alpha-decay, i.e. He. These gas atoms are often

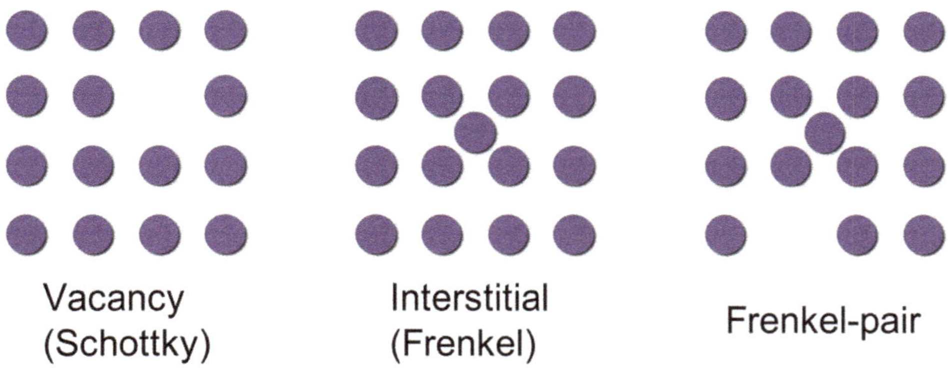

Figure 2.9. Examples of a Schottky (vacancy), Frenkel (interstitial) defects and a Frenkel pair.

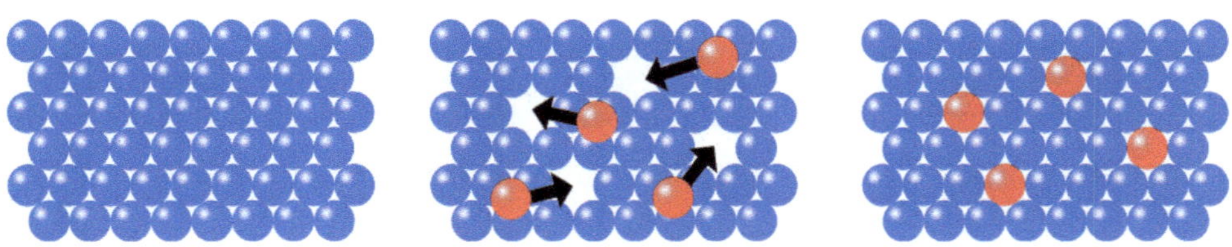

Figure 2.10. Idealistic schematic of interstitial/vacancy annihilation mechanism, see also animation.

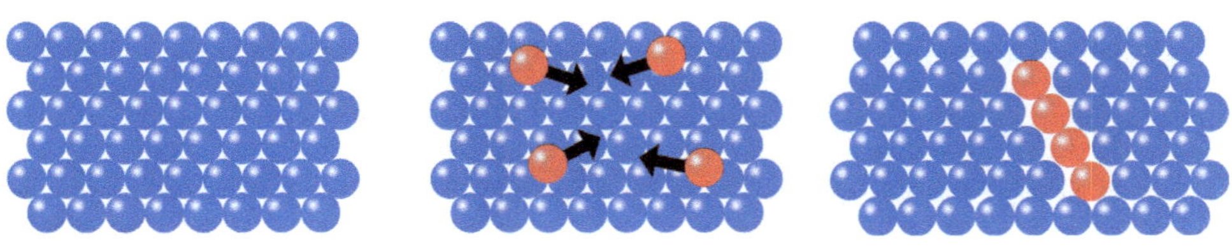

Figure 2.11. Formation of an interstitial defect loop from the migration of interstitial atoms, see also animation.

mobile in the lattice and can group together, leading to gas bubble formation in the structure (figure 2.13).

Thus for every damage cascade formed, there are multiple outcomes available for recovery, ranging from no recovery through to full recovery. Where there are multiple damage cascades within a core, it is highly probable that partial recovery will occur. It is how each of these simplistic recovery processes combines that will define the long-term stability of a material.

2.5.2 Damage versus recovery

For every induced damage cascade, multiple recovery profiles are available, dependent upon the material being irradiated, i.e. its composition, structure, temperature, etc. Each of these parameters impacts on the ability of the material to recover from damage, as in each case there will be different pathways available. These routes each have a different kinetic/thermodynamic nature, and which one is followed depends on many criteria; for the purposes of this discussion, we will assume that the most thermodynamically, or kinetically, favourable is followed, and that the higher the temperature, the more rapid the recovery. Referring to the examples shown in figure 2.14, it can be seen that the defects can move through the lattice and will tend to follow the easiest route available to them.

Up until now we have focused on recovery from a single cascade at a temperature, but if we now add a further parameter and include a further impact, two possibilities arise (figure 2.15):

(i) enhanced local recovery due to local temperature rise—this is where a partial recovery can continue because of a local increase in temperature from a neighbouring impact;

(ii) enhanced damage or recovery due to a subsequent pka passing through a partially damaged region of the lattice.

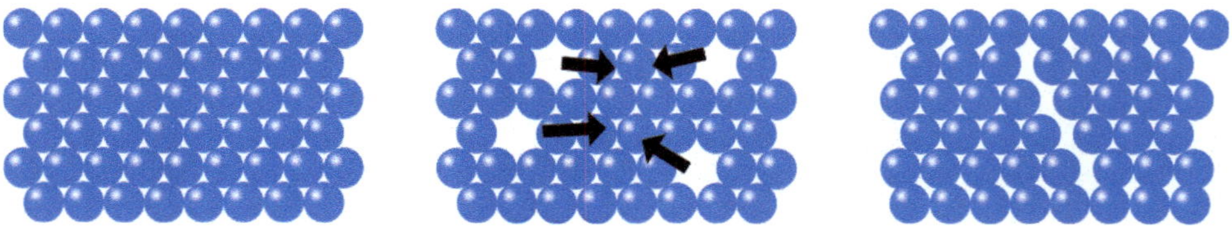

Figure 2.12. Formation of a vacancy defect loop from the migration of vacancies, see also animation.

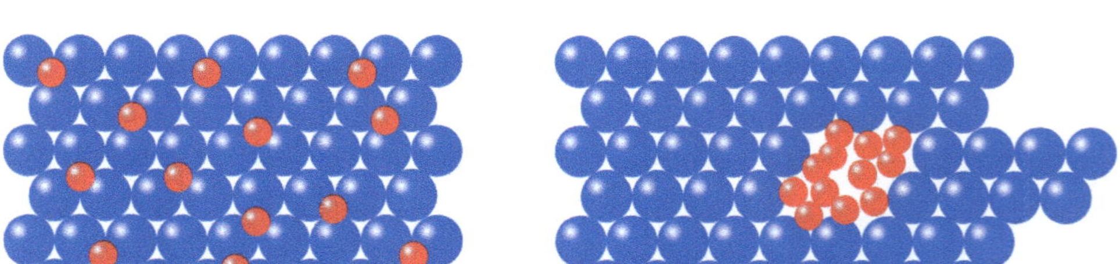

Figure 2.13. Formation of gas bubble from agglomeration of individual gas atoms, e.g He/Kr/Xe, see also animation.

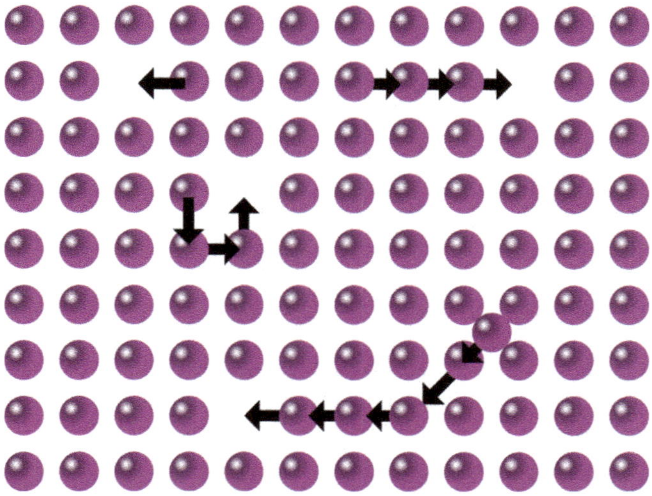

Figure 2.14. Options available for defects to migrate through a lattice, highlighting site swapping, linear vacancy migration, ring vacancy migration and vacancy annihilation, see also animation. These are just some of the mechanisms available.

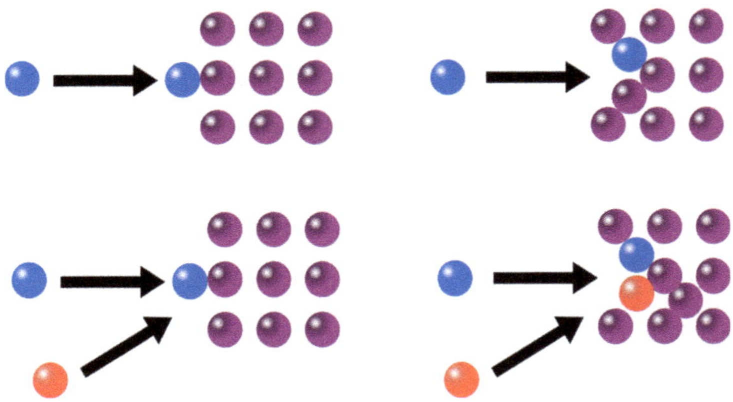

Figure 2.15. The two possibilities for enhanced recovery through further damage, see also animation.

Thus the overriding issue in developing new materials for use within a nuclear reactor, whether fission or fusion based, is how the materials will behave while undergoing competing processes, between damage and recovery, with multiple outcomes at each stage.

2.6 Modelling damage build-up in materials

How a material behaves with increasing damage, principally depends upon the volume of material immediately affected by the damage. Using two extremes in the range of damage, i.e. short or long range, we end up with two cases.

(i) The damage is localised in length, but impacts on a relatively large number of atoms of material locally. This type of damage process occurs during fission or alpha decay (in this case it is the recoiling atom), which we will describe as direct impact. An example of this is shown in figures 2.16 and 2.18.

(ii) The damage occurs over a much longer length scale, but may affect a much smaller volume of atoms, which we will describe as defect accumulation. An example of this is shown in figure 2.17.

2.6.1 Direct impact

This damage process arises from fission, or alpha decay, which has a large mass atom damaging a local volume of material. The splitting of the atom gives rise to the following:

i) a release in energy, binding energy release, increasing the local temperature of the material;

ii) a fast(er) moving particle of lower mass, e.g. He in the case of the alpha-decay of ^{239}Pu;

iii) a slow(er) moving particle of heavier mass, e.g. U in the case of the alpha-decay of ^{239}Pu.

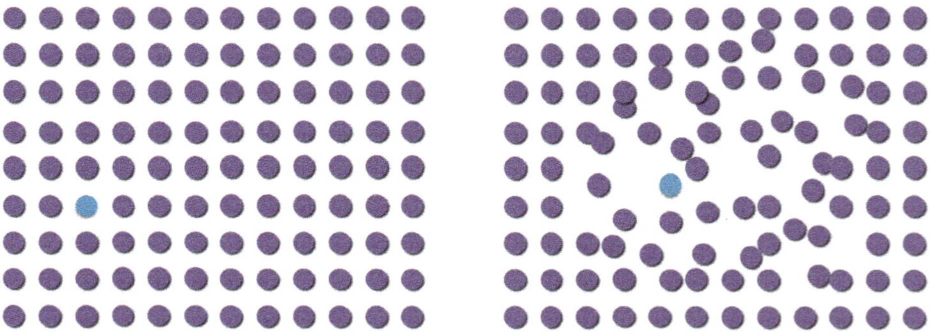

Figure 2.16. Example of direct impact damage, highlighting the small volume of material, but the large number of displacements arising from the recoil of the blue particle.

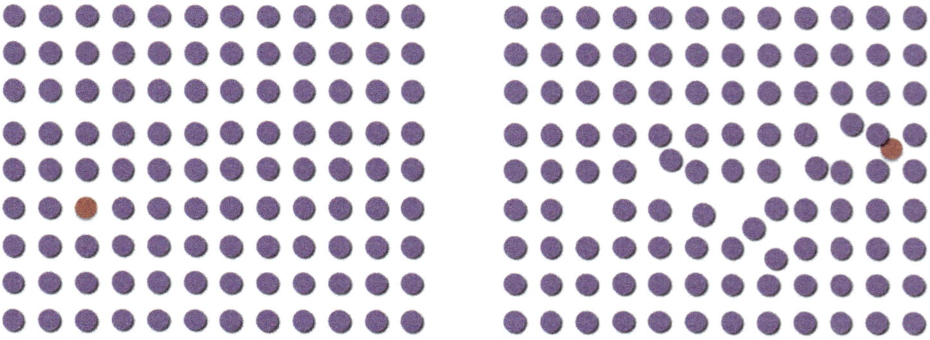

Figure 2.17. Example of defect accumulation showing a longer length of damage, but a smaller number of displacements arising from the impact of the red particle.

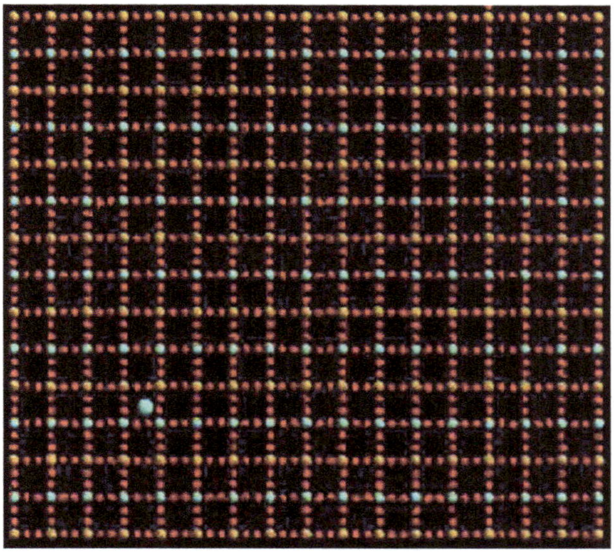

Figure 2.18. Cascade damage simulation within $ZrSiO_4$. Courtesy of Kostya Trachenko. See also animation.

Both of these particles give rise to damage, with the majority of the damage arising from the heavier particle, in a localised region.

2.6.2 Defect accumulation

This damage process arises from high energy–low mass particles, such as neutrons and He, which are assumed to follow the Kinchin–Pease methodology for describing damage. In essence, defect accumulation can induce damage over a relatively long range, but does not give rise to the same defect volume as direct impact, shown in figure 2.16.

2.6.3 How do they compare?

Direct impact will induce a large volume of damage very quickly, which then begins to recover back, but generally this process cannot be fully completed, i.e. there is partial recovery. Over time it has been found that when using natural minerals containing U, such as zircon, the fraction of amorphous material asymptotically reaches a level approximating to fully amorphous with increasing dose (figure 2.20).

Modelling this simplistically, the following relationship can be derived:

$$f_a = 1 - e^{(-\sigma_a D)}, \tag{2.10}$$

where σ_a is the amorphisation cross-section (the degree of amorphisation per damage event) and D is the total level of dose (the number of decaying atoms, for example). This is shown graphically in figure 2.19.

If this model is compared with the volume of amorphous material found in natural zircon, which has undergone significant damage from uranium decay, a close match is found (figure 2.20).

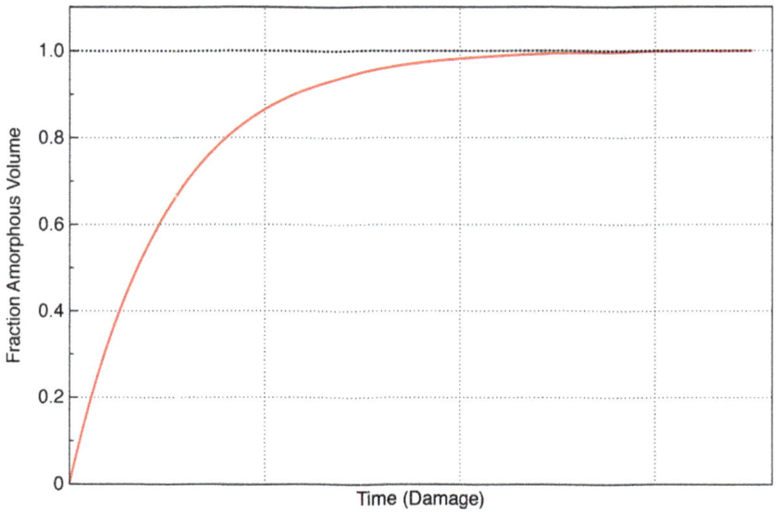

Figure 2.19. Graphical representation of equation (2.10).

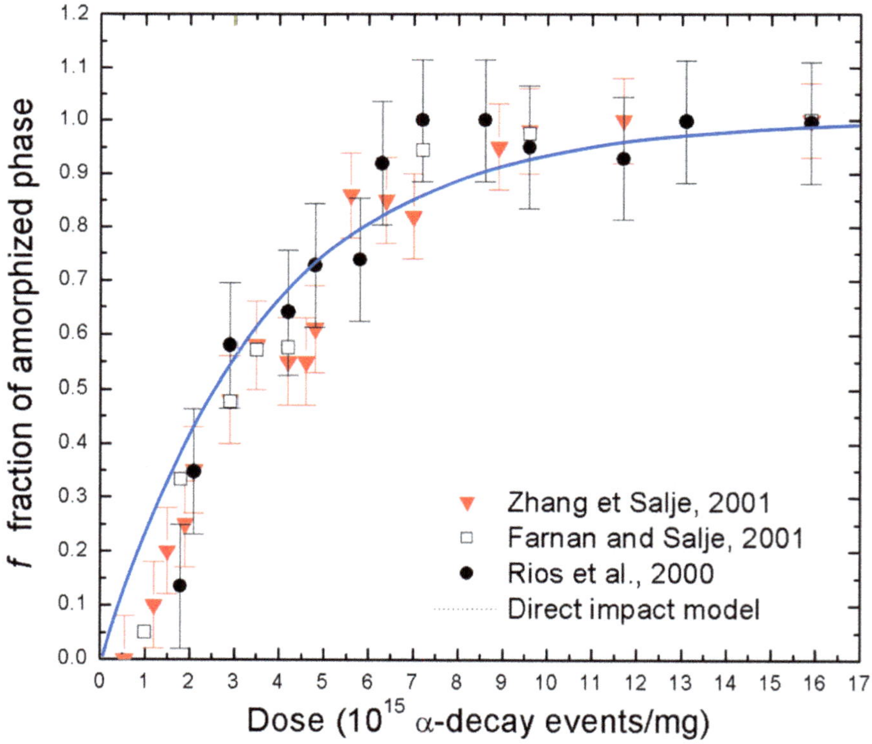

Figure 2.20. Fraction of amorphous phase within zircon as a function of damage, in this case defined as alpha-decays per mg of sample, with a direct impact model overlaid (blue line). Reproduced with permission from [7].

Defect accumulation (inducing damage over a longer range) allows the system to recover from damage more effectively, has a different formalism mathematically and is seen when materials are irradiated using an ion beam (e.g. figure 2.21). Mathematically, it is the inverse of the direct impact model, i.e.

$$f_a = e^{(\sigma_a D)} - 1, \quad (2.11)$$

where again σ_a is the amorphisation cross-section and D is the total dose (the flux of damaging particles, for example). This is shown graphically in figure 2.22.

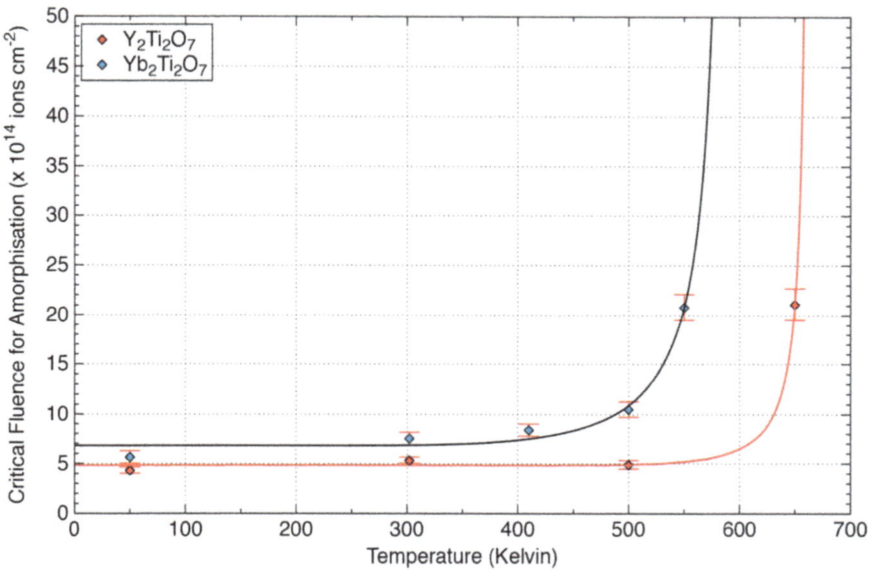

Figure 2.21. Fluence of ions required to amorphise the systems $Y_2Ti_2O_7$, and $Yb_2Ti_2O_7$ with 1 MeV Kr ions [5].

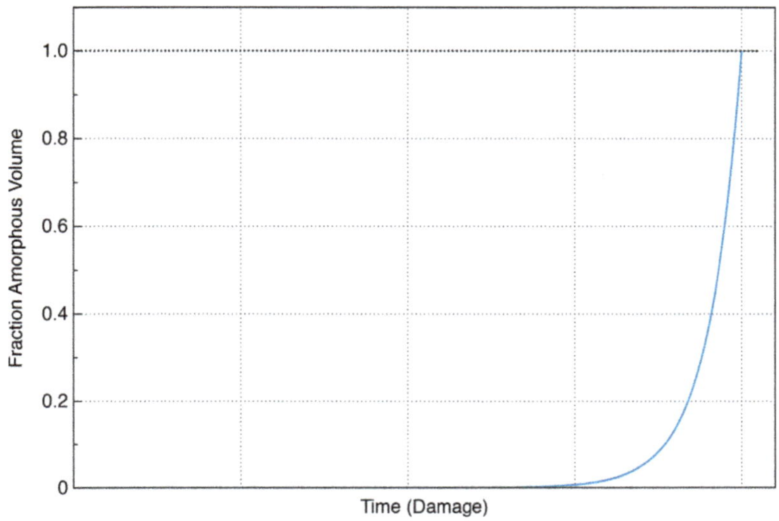

Figure 2.22. Graphical representation of equation (2.11).

When compared, it can be seen that they both reach the same level, i.e. 100% amorphisation, they just get there differently (figure 2.23). In reality, however, it is very rare for one model to occur exclusively, and a hybrid is observed (figure 2.24), which tends to be the case when heavy particles are inducing damage. At low levels of damage, a model based on defect accumulation often dominates, while at higher levels the direct impact model works better.

For example, using the data shown in figure 2.20 for zircon (from [7]), a hybrid model better describes how the damage builds up, as shown in figure 2.25.

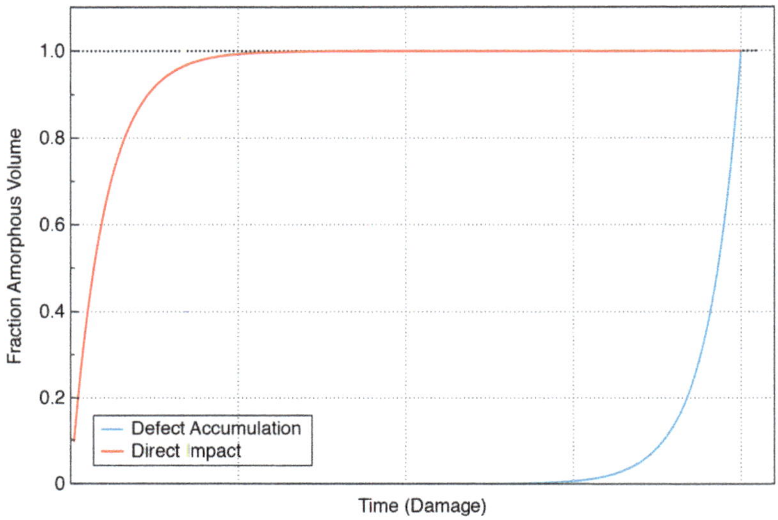

Figure 2.23. Comparison of the direct impact and defect accumulation models.

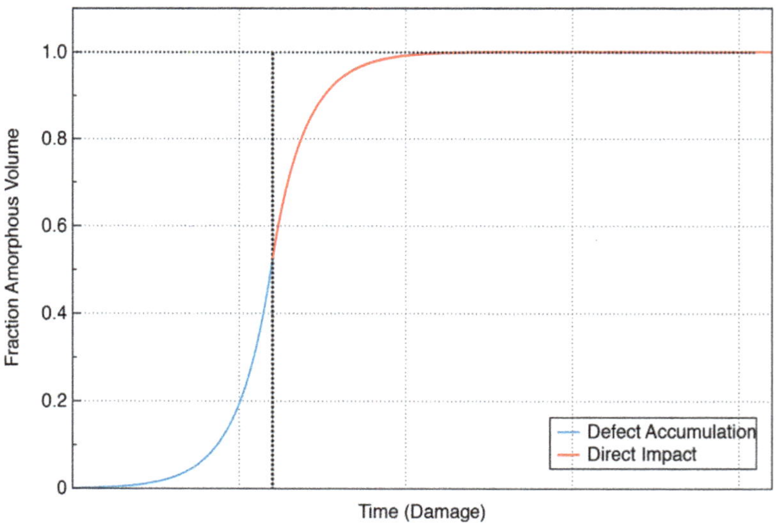

Figure 2.24. Hybrid of direct impact and defect accumulation models for amorphous fraction as a function of dose.

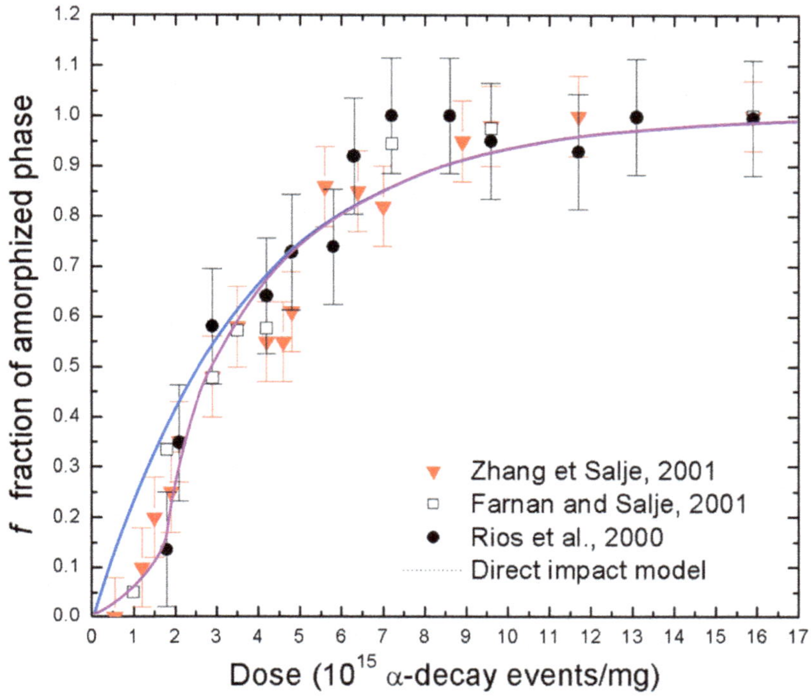

Figure 2.25. Repeat of figure 2.20 data, but with a hybrid model based on figure 2.24 overlaid, showing a better fit to the observed data (purple line) [6].

2.7 The bulk effects of damage

So far we have discussed how damage occurs, the typical end points for recovery from damage and how to model damage. However, we now need to discuss the key factor in radiation damage and materials. What are the bulk effects of these processes? This is important, as in essence this is the key outcome for any material. The key impacts from damage are multiple and each of them can prevent a material being used in a damaging radiation field, and in addition any amorphisation that occurs will often compound the problem.

2.7.1 Expansion (swelling)

Typically, when systems undergo damage, they will undergo volume expansion, which can arise from three mechanisms:

(i) vacancy and interstitial (defect) formation—generally found in simple systems, such as metals, it also increases the local strain/stress within a material;
(ii) decreased density due to damage—damaged systems by their very nature are rarely the most dense arrangement of atoms;
(iii) gas bubble formation—found in fuel, in systems where alpha decay occurs, or in plasma facing walls within fusion reactors, this links in with vacancy formation.

Amorphisation arising from a combination of the above processes can lead to excessive volume expansion, for example in a sample of 316SS considerable expansion was observed when it was irradiated to 1.5×10^{23} neutrons cm^{-2}, a not insignificant dose within a reactor core (figure 2.26).

One consequence of expansion is radiation-enhanced creep, where with increasing levels of damage, a material's length increases. For example, in Zircaloy-4 it has been seen that at 282 °C, the operating temperature within a boiling water reactor (BWR) core, there is significant expansion (figure 2.27).

2.7.2 Radiation-induced segregation

This can arise from a complex material, through the process of damage, separating into less complex simple materials. It has been observed in metallic alloys, where over time the alloy composition changes, rendering it unsuitable for use in a core.

2.7.3 Thermal conductivity

With increasing levels of damage, it is generally observed that thermal conductivity decreases, principally due to the lattice damage reducing the ability of the lattice to conduct heat through phonons. In figure 2.28, the effect of damage on thermal conductivity in SiC at different damage levels is shown, and as can be seen, with increasing damage, the thermal conductivity can drop dramatically.

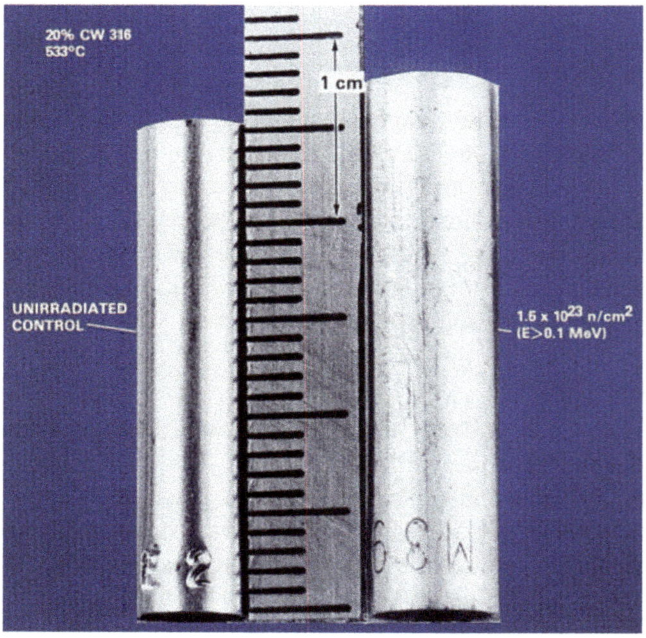

Figure 2.26. Example of volume swelling in 316 steel irradiated with 1.5×10^{23} n cm^{-2} at 533° C [7].

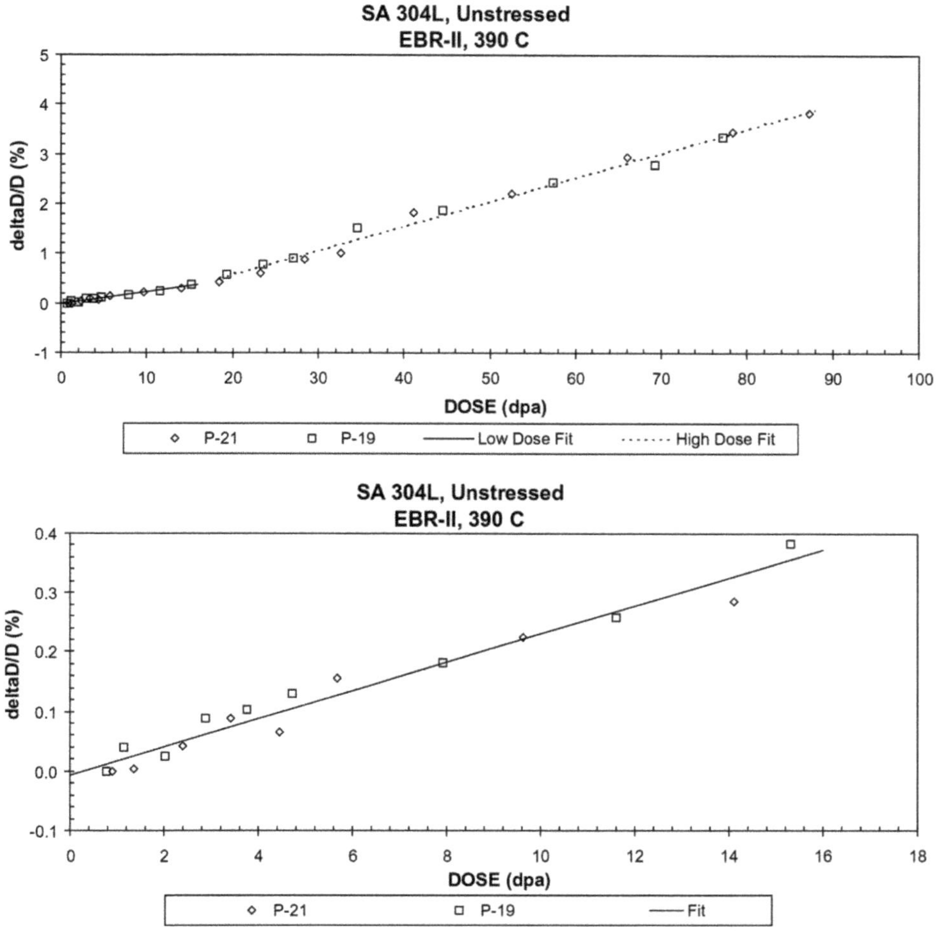

Figure 2.27. Example of length change (creep) in Zircaloy-4 when irradiated with neutrons at ~282 °C [8].

2.7.4 Embrittlement

Damage that induces changes in the material can also lead to it losing some ductility and becoming more brittle. In many cases, while the ultimate tensile stress increases, it occurs at a much lower strain. This can directly prevent a material being used in a core, as it no longer meets the requirement for ductility after irradiation shown in figure 2.29.

2.7.5 Cracking

Cracking occurs when the material can no longer accommodate the induced stress/strain energy and it relieves this excess energy through the formation of a crack, examples shown in figure 2.30. Such cracks can occur throughout the material, but they often initiate at defects induced by damage, i.e. vacancies/interstitials.

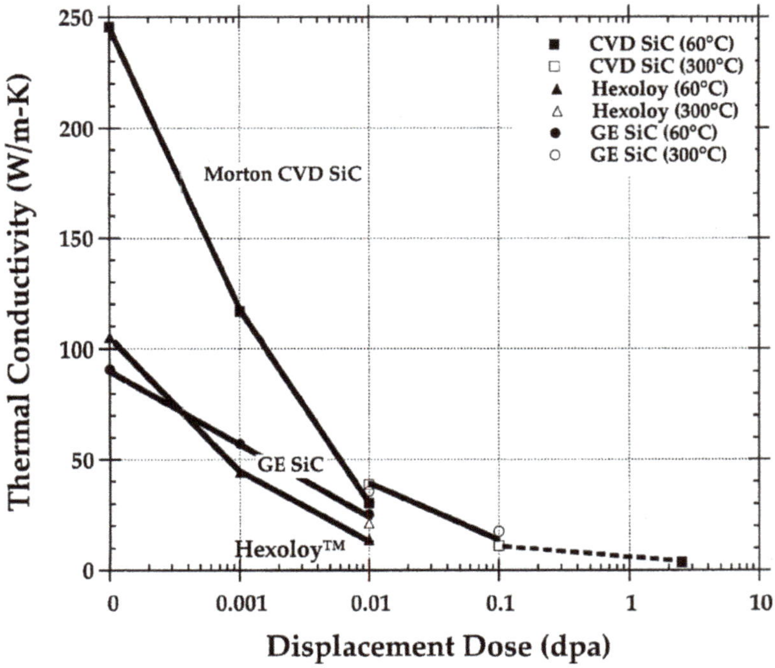

Figure 2.28. Degradation in thermal conductivity in SiC with increasing levels of damage [9].

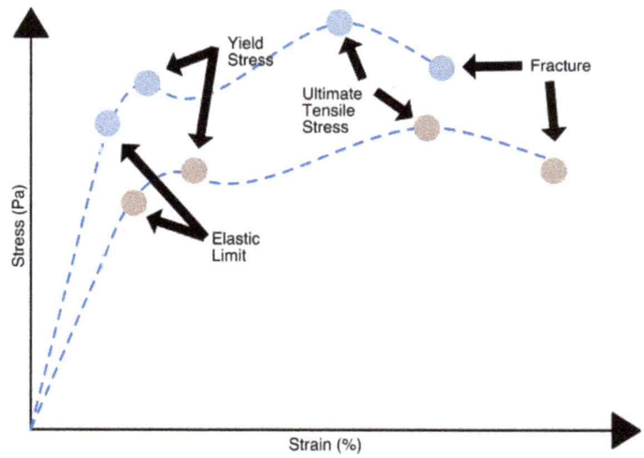

Figure 2.29. Schematic representation of changes in stress as a function of strain in irradiated and unirradiated materials.

Each of the above effects on their own can render a material unusable within a high radiation field, e.g. a core. In general, there are materials capable of being used in reactor cores, designed to experience ~60 dpa, the challenge going forward is how to design materials for the next generation of reactors.

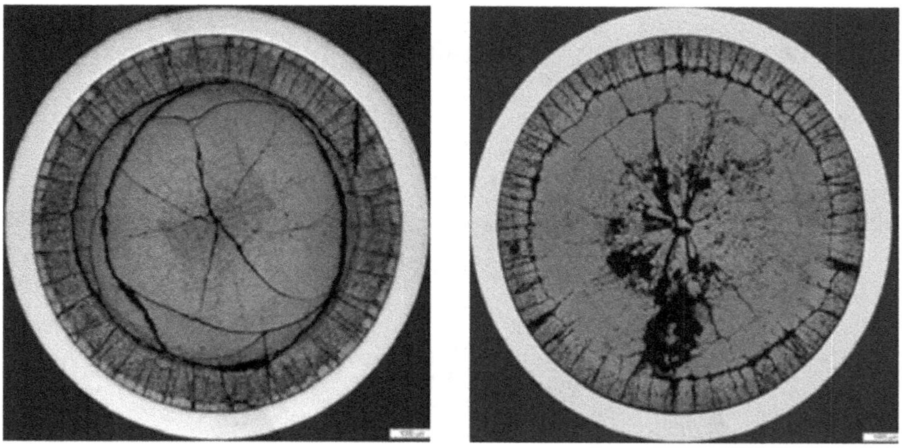

Figure 2.30. Cracks in nuclear fuel formed partially through induced radiation damage [10].

One facet of radiation damage that has not been discussed so far is the effects on a material where we know there is significant damage, but where we have no alternative. Such an example is nuclear fuel, which we will discuss in the next chapter.

References

[1] Kinchin G H and Pease R S 1955 *Rep. Prog. Phys.* **18** 1–51
[2] Ziegler J F, Ziegler M D and Biersak J P 2010 *Nucl. Instr. Meth. Phys. Res.* B **268** 1818–23
[3] Norgett M J, Robinson M T and Torrens I M 1975 *Nucl. Eng. Des.* **33** 50–54
[4] ASTM 2012 E693 Standard practice for characterizing neutron exposures in ferritic steels in terms of displacements per atom (dpa) *Annual Book of ASTM Standards* vol 12.02 (West Conshohocken, PA: ASTM International)
[5] Whittle K R, Blackford M G, Aughterson R D, Lumpkin G R and Zaluzec N J 2011 *Acta Materialia* **59** 7530–7
[6] Rios S and Salje E K H 1999 *J. Phys.: Condens. Matter* **11** 8947
[7] Straalsund J L, Powell R W and Chin B A 1982 An overview of neutron irradiation effects in LMFBR materials *J. Nucl. Mater.* **108/109** 299-305
[8] Foster J P, Bunde K and Porter D L 2003 *J. Nucl. Mater.* **317** 167–74
[9] Snead L L, Zinkle S J and White D P 2005 *J. Nucl. Mater.* **340** 187–202
[10] IAEA 2010 Advanced fuel pellet materials and fuel rod design for water cooled reactors *IAEA TECDOC* 1654

Karl Whittle

Chapter 3

Nuclear fuel, part 1: fuel and cladding

The key requirement in any nuclear reactor, whether it be fission or fusion based, is fuel, and this chapter focuses on introducing fuel and cladding for fission-based reactors. Before discussing the various options, a review of the key requirements for nuclear fuel, and how it links into the fission process itself, is apt.

3.1 What is required from fuel in a fission reactor?

The essential requirement of fuel is that it must contain a source of fissile or fertile material. For example, fuel should contain a fissile isotope, such as ^{235}U, ^{239}Pu or ^{233}U, or, if a fertile (breeding) reactor is being used, ^{238}U and $_{232}Th$ can be employed. Without either of these, the reactor cannot induce fission, and therefore cannot be used to generate electricity.

What is often equally important, and is a driving factor in reactor design, is the ability of the fuel to be stable and conduct heat away. Remember that a reactor uses the heat generated from fission to heat water to steam, or gas, which then drives turbines to generate electricity. Therefore, having good thermal conductivity is essential.

The final key requirement is fuel stability, not only to the temperature increase, but also to chemical changes, through the transmutation from fission and induced radiation damage from the energetic particles formed.

3.2 Reminder of the fission process

Remember that during fission (figure 3.1) the end product is heat (which we use to indirectly generate electricity) and multiple particles, i.e. fission products and neutrons. This is not a trivial process, it generates large amounts of energy, but, as was seen in chapter 2, significant amounts of damage. These are what fuel has to be as resistant as possible to, while at the same time allowing the fission process to proceed, i.e. low neutron absorption and heat escape to the cooling medium are required. This is not a trivial problem to overcome.

doi:10.1088/978-0-7503-1104-5ch3

3.3 What are the realistic types of fuel?

In principle, it would be possible to use solid, liquid or gaseous fuel, assuming the conditions were correct. However, realistically, it is only possible to use either solid or liquid fuel, and in most reactors operating today only solid fuel is used. Even when using solid fuel, there is more than one type available, for example it could be metallic or ceramic. However, the most common form of fuel used today is ceramic, based on UO_2, but before we discuss UO_2, we shall consider the early fuels, which were based on uranium metal, as this provides an insight into why UO_2 is now more popular.

3.4 Uranium

Uranium, as any chemist will tell you, is the heaviest of the naturally occurring terrestrial elements, i.e. those found on Earth. However, this does not mean it was the heaviest naturally occurring element at the formation of the Earth, if you compare the half-lives of the heavy elements, uranium just happens to be the one with the longer half-lives, and is thus still present on the planet. Uranium is found naturally with three isotopes, as shown in table 3.1.

As can be seen, each of the isotopes undergoes decay, ^{234}U is itself a decay product arising from the decay of ^{238}U via ^{234}Th and ^{234}Pa.

Uranium as an element has a range of oxidation states, arising from its electronic structure, *[Rn]* $7s^2$ $6d^1$ $5f^3$. This gives rise to six potential oxidation states, 1–6, however, the most common ones are 4 and 6—these oxidation states will become more important in the next section on oxides. As with most metals, uranium is silvery/white

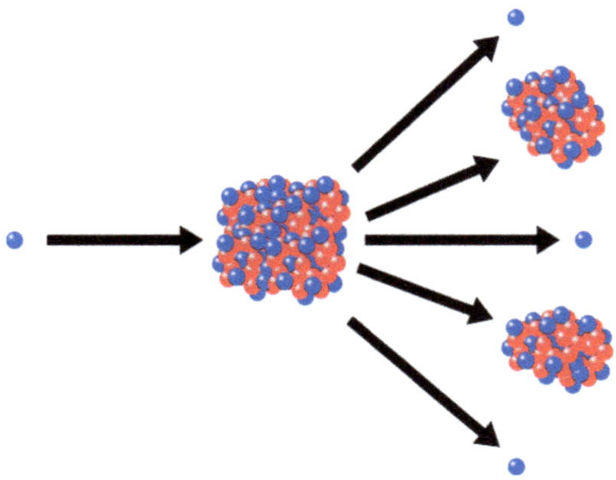

Figure 3.1. Induced fission and the products, see also animation.

Table 3.1. Natural abundance of uranium isotopes, their fission/fertile nature and what they become if fertile [1].

Isotope	Half-life (yr)	Relative amount (%)	Fissile/fertile
^{234}U	2.46×10^5	0.0054	Fertile (^{235}U)
^{235}U	7.04×10^8	0.7204	Fissile
^{238}U	4.47×10^9	99.2742	Fertile (^{239}Pu)

in nature, electrically conductive, has a good thermal conductivity of 27.5 W m^{-1} K^{-1} and, as would be expected, a high density at room temperature of 19.1 kg m^{-3}. The key point to remember about density is that it has implications for use within a reactor core, as uranium undergoes two phase changes with temperature, modifying its density.

At temperatures below 935 K, it is orthorhombic in nature, between 935 and 1045 K, it is tetragonal, whilst above 1045 K, it is cubic, until it melts at 1405 K crystal structures shown in figure 3.2, phase diagram in figure 3.3. Why is this important?

These changes are important for multiple reasons, however, two things in particular are important for its use as fuel, the changes in crystal structure and the relatively low melting point. As the fuel heats up during operation it will change structure, modifying its density and expanding, which could be problematic in use. In addition, the melting

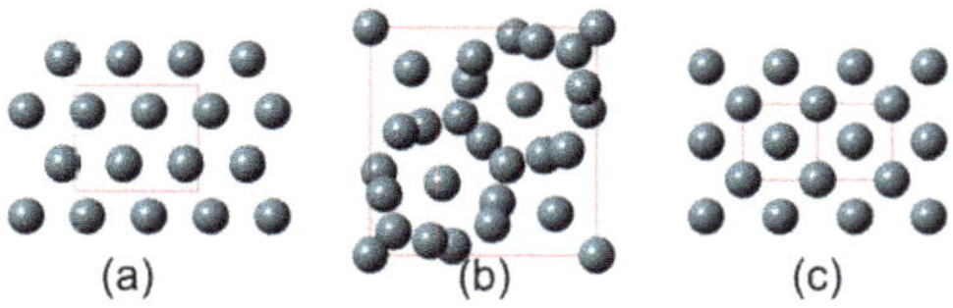

Figure 3.2. Crystal structures of uranium: (a) orthorhombic, (b) tetragonal and (c) cubic [2].

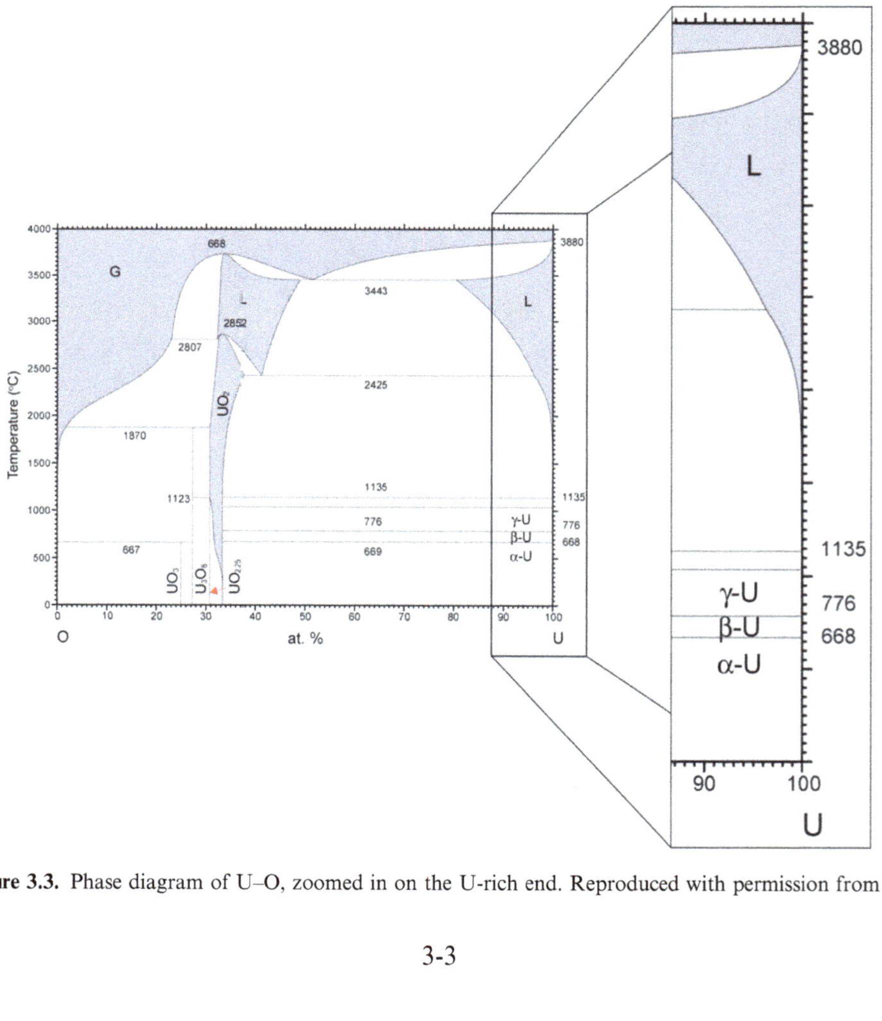

Figure 3.3. Phase diagram of U–O, zoomed in on the U-rich end. Reproduced with permission from [3].

point is not ideal. Since the fission process can easily create local temperatures in excess of 1500 K, having a fuel melt, or even soften, would cause issues.

As outlined in the beginning of this chapter, one of the key criteria for fuel, outside it containing fissile/fertile material, is chemical and thermal stability. Uranium as a metal is by its nature not the most chemically stable form of uranium, but the oxides are, with much higher melting points, giving rise to further stability and the potential for higher temperature of operation.

3.4.1 Oxides of uranium

The most stable oxidation states for uranium are U^{4+} and U^{6+}, arising from the orbital structure *[Rn]* $7s^2\ 6d^1\ 5f^3$. Mineral sources for uranium are predominantly oxides such as uraninite (UO_2) and pitchblende (UO_2/UO_3), with UO_2 forming the basis of currently used nuclear fuel.

3.4.2 Uranium(IV) oxide (UO_2)

Uranium(IV) oxide adopts a cubic crystal structure (figure 3.4) similar to CaF_2, i.e. fluorite, with a melting point of 3138 K, significantly higher than that of metallic U, but a density of 10.97 kg m^{-3}, which is less than that of U. The one big advantage UO_2 has over metallic U is its cubic nature, with no phase transitions until melting, and therefore a predictable isotropic thermal expansion.

The thermal expansion of UO_2 can be approximated to linearity below ~1700 K, but above this temperature it becomes non-linear (figure 3.5) [4]:

$$\frac{L}{L_{273}} = 0.99672 + (1.179 \times 10^{-5})T - (2.429 \times 10^{-9})T^2 + (1.219 \times 10^{-12})T^3. \quad (3.1)$$

The thermal conductivity of UO_2, similar to most ceramics, decreases with increasing temperature, but begins to increase above ~2200 K (figure 3.6) [5].

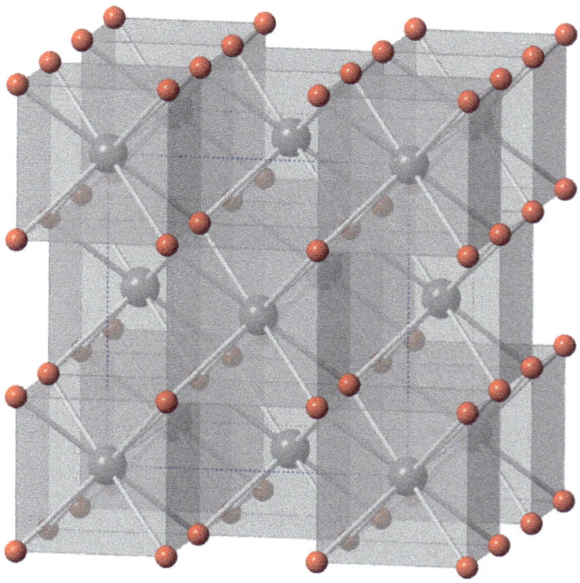

Figure 3.4. Crystal structure of UO_2.

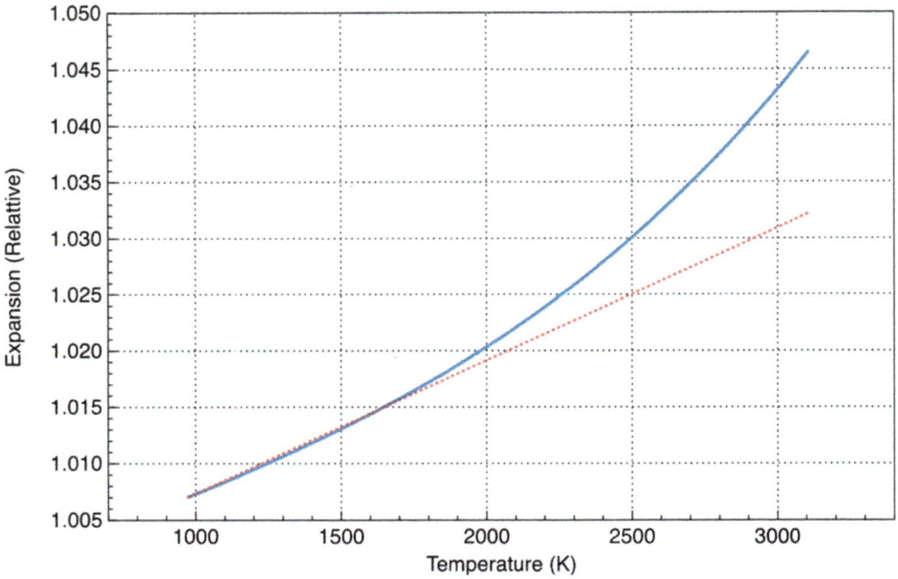

Figure 3.5. Thermal expansion within UO_2 as a function of temperature, the red line indicates linearity.

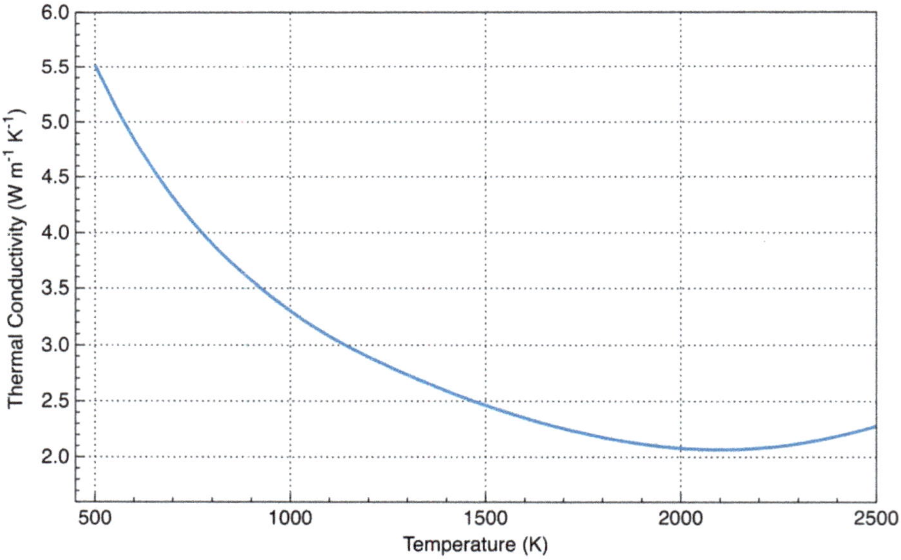

Figure 3.6. Thermal conductivity of UO_2 as a function of temperature.

3.4.3 Uranium (VI) oxide (UO_3)

As outlined above, UO_2 is just one of the oxides available for uranium, the other common one is UO_3, which has different properties. For example, whereas UO_2 adopts a cubic structure based on fluorite, UO_3 has four possible crystal structures, with differing symmetries, from cubic through to monoclinic. As a result, the melting point can vary, but is below 923 K, a temperature that would cause problems similar to those experienced when using uranium metal, if it was used as fuel.

3.4.4 Uranium (IV/VI) oxides

It is possible, and often found, that uranium will form mixed oxides based on a mixture of U(IV) and U(VI). The two common ones are U_4O_9 and U_3O_8, each having different properties and potential uses:

- U_4O_9—this system is a mix of $U^{6+}U_3^{4+}O_9$, adopting a structure similar to that of UO_2
- U_3O_8—this is a mix of $U_2^{6+}U^{4+}O_8$, adopting an orthorhombic structure.

Each of these oxides is non-ideal for use within fuel, but they have advantages in the transportation of material, for example U_3O_8 is used to transport enriched fuel, prior to fabrication into pellets. If these oxides are combined with uranium, and the phase diagram for uranium–oxygen is examined, it can be seen that it is nowhere near as simple, and indeed is highly complex. The key region is highlighted in figure 3.7.

Taking thermal and chemical stability into account, the ideal form of U–O to use is UO_2. However, it is not quite as simple as that, generally UO_2 is slightly non-stoichiometric and found as $UO_{2\pm x}$.

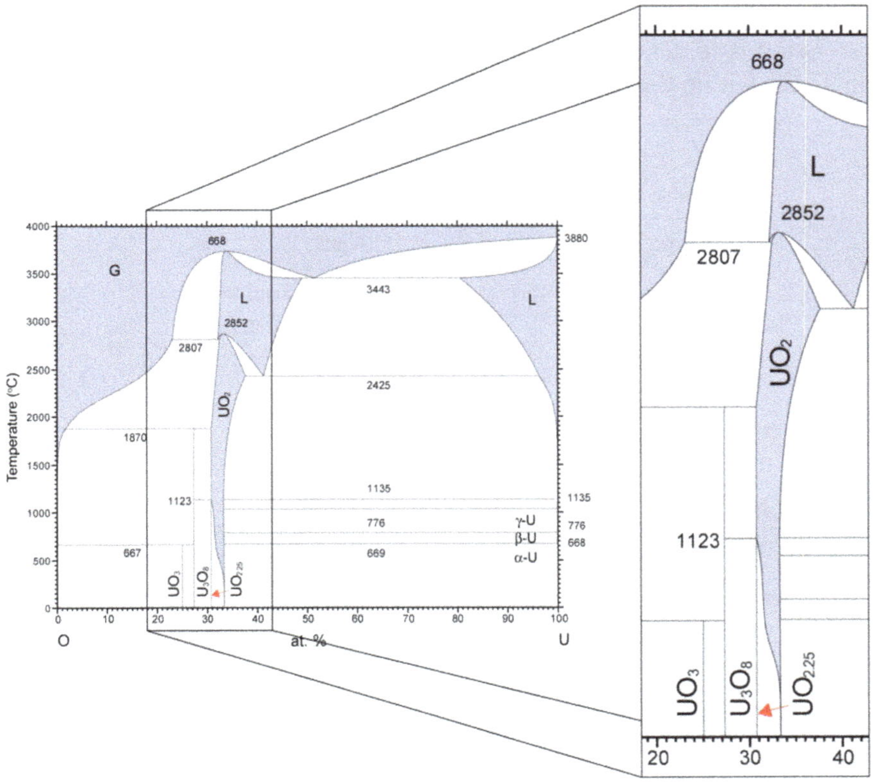

Figure 3.7. The phase diagram of U–O, with the region around UO_2–UO_3 highlighted [3].

3.4.5 Hyper/hypostoichimetric $UO_{2\pm x}$

The slight changes in U–O ratio around UO_2 can have profound effects on the properties of UO_2 and give rise to issues during use. Before we discuss such effects, we need to define what hyper- and hypostochiometric are, and they are summarised in table 3.2.

Hypostoichiometic UO_{2-x} has partial reduction of U(IV), with oxygen vacancies to maintain charge balance. Hyperstoichiometric UO_{2+x} contains additional oxygen atoms in the oxygen array (these are located within specific crystallographic positions), with partial oxidation of U(IV) to U(VI). The extra oxygen required for charge compensation is found to locate on specific locations within the lattice, as highlighted in figure 3.8.

Each of these cases has a profound impact on the ability of UO_2 to be used as fuel, primarily due to the effects on thermal conductivity. As is well known, thermal conductivity in materials is through lattice phonons, which are impacted on by both structure and composition. If we consider that UO_2 is a perfect lattice, which is a reasonable assumption to make, then as vacancies/interstitials are added, the lattice is no longer perfect, impacting on the conductivity.

As can be seen in figure 3.9, as the composition moves away from UO_2, the conductivity drops dramatically. As an aside, U_4O_9, the first stoichiometric oxide above UO_2, is $UO_{2.25}$, and therefore expected to have lower conductivity. Why are we worried about poor thermal conductivity? As you may recall, the role of nuclear fission is to generate heat, which is then converted to electricity. If we are retaining heat within the fuel, then the process is fundamentally inefficient, but that

Table 3.2. Definitions of hyper/hypostoichiometry.

Type	Name
UO_{2-x}	**Hypo**stoichiometric
UO_2	Stoichiometric
UO_{2+x}	**Hyper**stoichiometrc

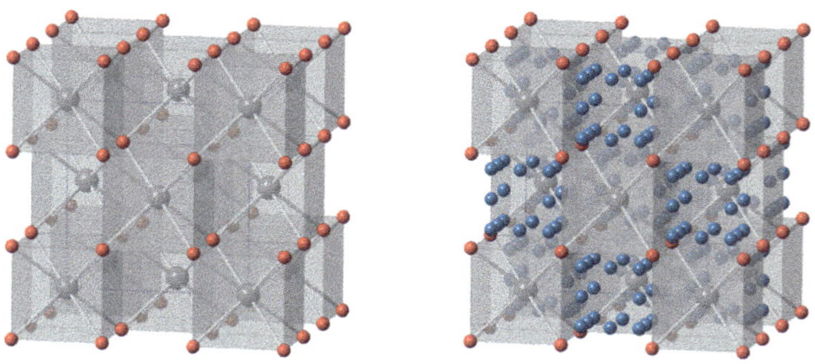

Figure 3.8. Comparison of UO_2 and UO_{2+x}. The extra potential oxygen atom locations are highlighted in blue [6].

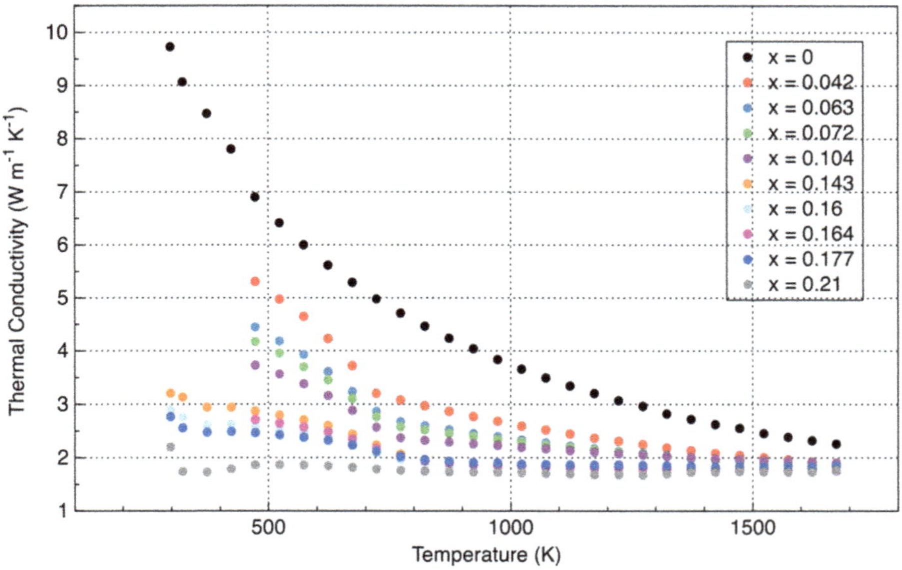

Figure 3.9. Thermal conductivity as a function of *x* in $UO_{2\pm x}$ [7].

is not the main concern. What is of critical concern is that heat is retained within the fuel, leading to melting. There are, however, methods available by which the thermal conductivity can be increased, for example adding additives when fabricating the pellets, such as TiO_2 or metallic additives.

So far, we have discussed one potential fuel source, uranium, and one other element to consider is plutonium.

3.5 Plutonium

Unlike uranium, plutonium is not naturally found terrestrially. It is likely that it was found in the past, but due to its short half-life compared with uranium, it is unlikely to be found now. Plutonium, like uranium, is metallic, and has a range of oxidation states arising from its electronic structure, in this case *[Rn] 7s2 5f6*, which is markedly different from uranium's, *[Rn] 7s2 6d1 5f3*. Whereas in uranium there are six oxidation states, with two common ones, plutonium has eight potential states, with Pu(III) and Pu(IV) being the ones of interest for nuclear fuel.

Plutonium in the metallic state, when compared with uranium, has a slightly higher density of 19.8 kg m^{-3}, but poorer thermal conductivity of 6.74 W m^{-1} K^{-1}, and a lower melting point of 913 K. Similarly, when compared to uranium, it undergoes more phase transitions when heated; whereas uranium has two phase transitions, plutonium has five. As a result, Pu metal is not ideal for use as fuel without stabilisers.

3.5.1 Oxides of plutonium

As with uranium, plutonium can adopt many different oxide compositions; the commonly found ones are based on Pu(III) and Pu(IV) oxidation states. For example,

Pu_2O_3 adopts a hexagonal structure, and similarly to UO_2, can also be hyper-stoichiometric and form $PuO_{1.52}$ and $PuO_{1.61}$. However, the oxide of plutonium with the most interest for nuclear fuel is PuO_2, which has Pu^{+4} oxidation state.

3.5.2 Plutonium (IV) oxide (PuO_2)

This oxide adopts the same structure type as UO_2, i.e. fluorite, with a similar density of 11.6 kg m^{-3}, and a melting point of 2670 K, shown in figure 3.10. These comparable properties are shown in table 3.3 and highlight the close match in properties.

As can be seen in table 3.3, the properties are very similar, but does this mean we can mix the two systems? Obviously, the answer is yes, otherwise this section would come to a very quick end. The system we end up with is known as mixed oxides (MOX), and these have been proposed, and indeed used, in many light water reactors (LWR). Bearing in mind that we have mentioned that moving from an ideal modifies thermal conductivity, does mixing the oxides modify the properties, and are there any caveats we should be aware of?

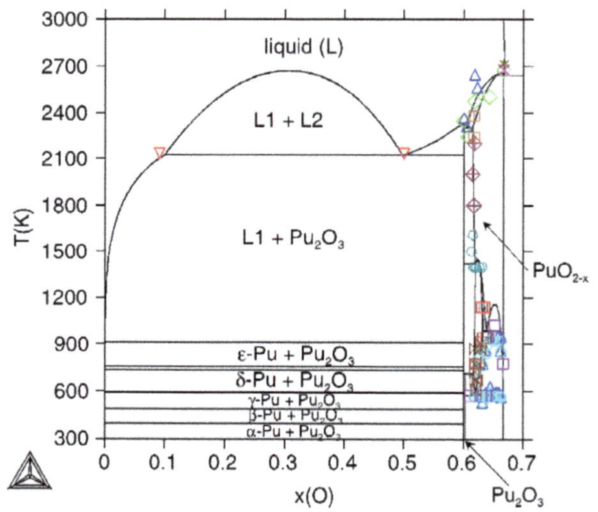

Figure 3.10. Phase diagram of Pu–O, focusing on the Pu-rich end and highlighting the phase transitions with increasing temperature [8].

Table 3.3. Summary of physical parameters for UO_2 and PuO_2 [1].

Property	UO_2	PuO_2
Melting point (K)	3138	2670
Density (kg m^{-3})	10.97	11.6
Metal ionic radius (angstrom)	1	0.96
Crystal system	Cubic	Cubic
Lattice parameter (angstrom)	5.468	5.398

3.5.3 Mixed oxide fuel (MOX)

Luckily, when mixing, since both systems are similar, there are no major problems, for example they form a solid solution across the whole compositional range, with no miscibility gap or separation. This solid solution gives rise to predictable lattice parameters, and while there is a reduction in the melting point, with a concomitant increase in the boiling point, in the middle of the range these changes do not prevent its use as fuel (figure 3.11).

If you remember when were discussing the effect of hyper/hypostoichiometry on thermal conductivity, there was a concern that modifying the composition, introducing defects, impacts on conductivity negatively. In MOX the situation is still prevalent, but as with most things in the nuclear field, it is something we can live with if we need to. In the case of MOX, as shown in figure 3.12, the effect is initially detrimental at low temperatures, but with increasing temperature it becomes less of a problem.

So why would we use MOX fuel, when we could just as easily use UO_2 or PuO_2? The answer can be broken down into two parts:

(i) it provides a basis by which excess Pu inventory can be used within current reactor technologies, and
(ii) it can act as a seed for fertile fuels, such as ^{238}U, which can then be used to extend the available fuel further.

3.5.4 What form is the fuel?

In general, most fuel UO_2/MOX-based fuel is fabricated into pellets that are ~1 cm in diameter and height, and these are then stacked with tubes, as shown in figure 3.13. There are two types of fuel pellet, one has an annulus in the middle to

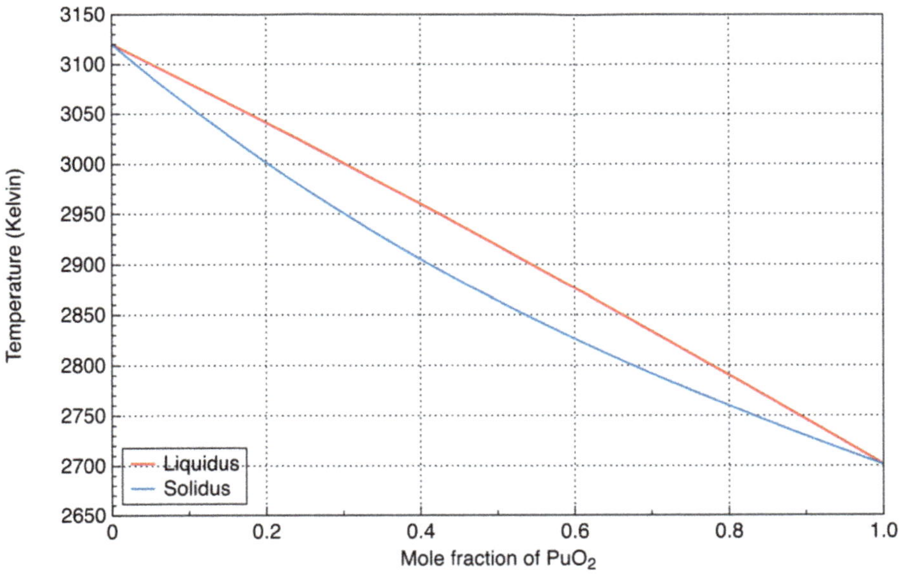

Figure 3.11. The modification of the melting point with increasing Pu content [9].

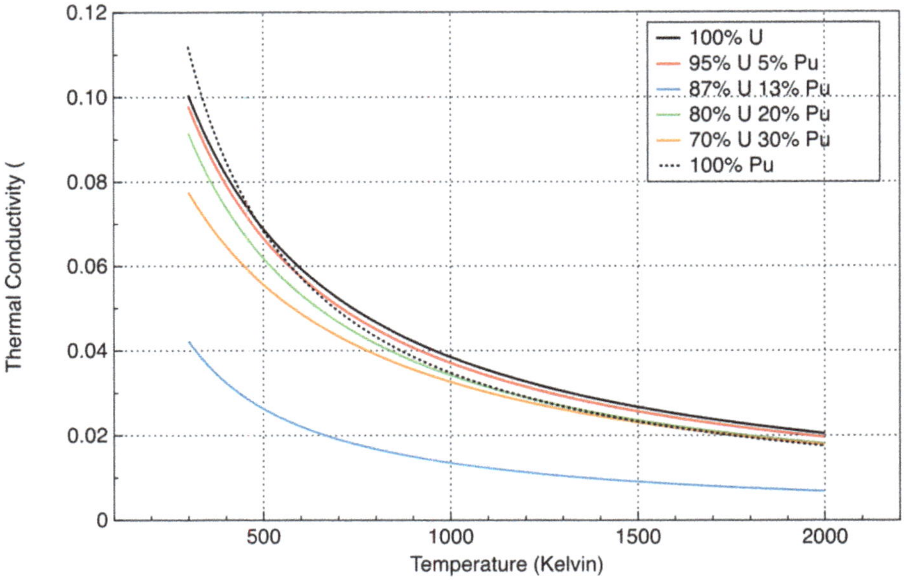

Figure 3.12. Thermal conductivity of UO_2, PuO_2, $(U,Pu)O_2$ and $U_{0.9}Gd_{0.1}O_2$ as a function of temperature, highlighting the loss in conductivity at low temperature [10].

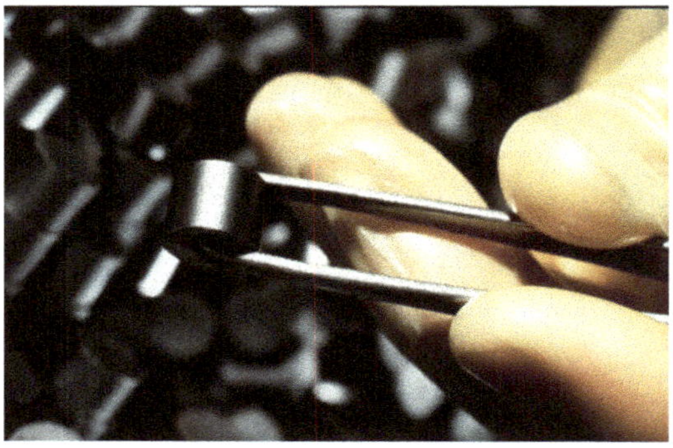

Figure 3.13. Image of a fuel pellet before irradiation within a reactor. Reproduced with permission from [11].

aid with fission gas release, and the other is solid. We will discuss how the fuel pellet behaves in a reactor core later, but first we must discuss the methods available for containment of the fuel pellets.

3.6 Fuel containment

While UO_2 may be relatively stable, under oxidising conditions it slowly transforms into UO_3, which, as you may recall, has detrimental properties when used in a reactor, i.e. a low melting point, etc. To compensate for this potential, and also provide a structure for the fuel, fuel generally does not come into contact with the coolant being used. The fuel is clad in a non-reactive material, typically Fe or Zr based.

In a manner similar to fuel, cladding has three essential requirements, and two of these are similar to those for fuel, good chemical stability and thermal conductivity. However, for cladding there is a third criterion, which is implied in fuel, but required in cladding, the ability for neutrons to pass through with low absorption, i.e. neutron transport. This transparency, allowing neutrons to pass through from one fuel rod to another and into the moderator, increases both fission efficiency and use of fuel. In an ideal world we would use the lowest neutron absorption material as cladding, however, we can only use cladding that also meets the other two criteria, i.e. chemical stability and thermal conductivity, or since metals are good thermal conductors, a material with an acceptable melting point. Of course, since we are trying to make the process as economically efficient as possible, we would also like to choose a material that is relatively cheap to manufacture and use.

When trying to decide which candidates can be used for cladding, let us set some criteria for their selection, based on the requirements for clad:

(i) melting point—must not be too low, to allow for efficient reactor operation;
(ii) chemical compatibility—must not be a highly reactive element, i.e. must be chemically inert with both the fuel and coolant used;
(iii) manufacturable—can it be formed into the required shape/geometry;
(iv) mechanically stable—it is required to contain fuel, so therefore should not undergo major changes within the reactor core (later we will discuss some of the causes of cladding changes);
(v) acceptable neutron absorption—while it is ideal to have the lowest possible, it may not be practicable, for example the lowest neutron absorption for any naturally occurring element is seen in oxygen, and we obviously cannot make cladding out of that.

So, using the melting point and acceptable neutron absorption as an initial metric, and plotting them against each other, we can now begin to choose potential candidates (figure 3.14).

From close examination of figure 3.14, some potential cladding materials could be based on the elements show in table 3.4, where the melting point and neutron absorption are compared.

As can be seen, the ideal candidate for cladding would be carbon, and indeed it is used in nuclear reactors, but predominantly as a moderator, i.e. CO_2 within an advanced gas-cooled reactor (AGR), or for structural purposes, again within an AGR. Carbon has potential problems in that its properties are not ideal for either manufacturing or mechanical stability, it has very little ductility and it can be very brittle. Beryllium has also been tried as cladding, but as with carbon, its ductility does not make it conducive for use within a reactor.

The early cladding, for example as used in the Hanford reactors, was based around the use of aluminium, in this case uranium metallic fuel contained with aluminium cans, within aluminium tubes. An alternative approach taken in the UK, within the MAGNOX family of reactors, was to use a magnesium–aluminium alloy (for example containing, Al 0.08 wt% and Be 0.004 wt%) [12]. These reactors, as

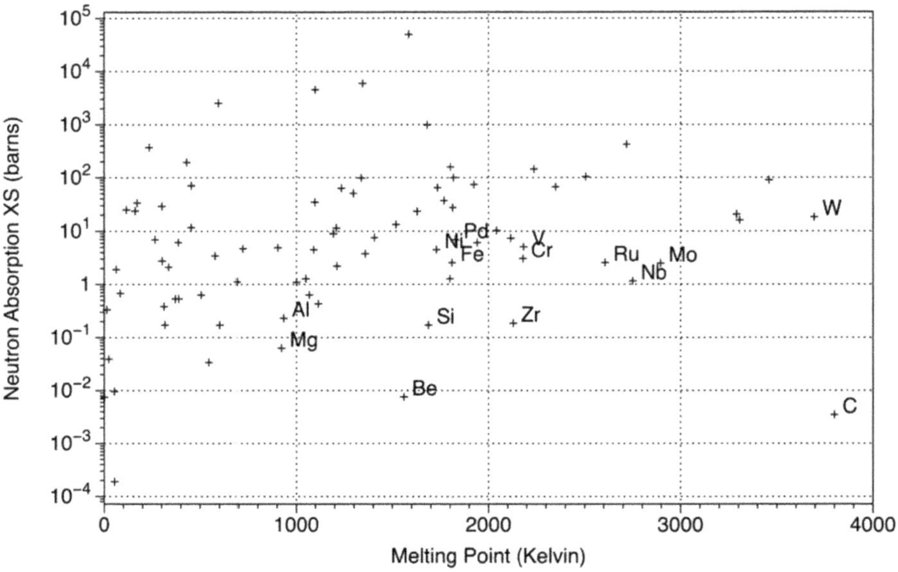

Figure 3.14. Plot of neutron absorption *XS* against melting point for the naturally occurring elements. The highlighted elements are those of interest for use within a fission reactor. Note the logarithmic scale. Data taken from [1].

Table 3.4. Selected natural neutron absorption cross-sections and melting points, sorted by increasing absorption [1].

Element	Abs *XS* (barns)	Melting point (kelvin)
C	0.0035	3800
Be	0.0076	1560
Mg	0.063	923
Si	0.171	1687
Zr	0.185	2128
Al	0.231	933
Nb	1.15	2750
Mo	2.48	2896
Fe	2.56	1811
Ru	2.56	2607
Cr	3.05	2180
Ni	4.49	1728
V	5.08	2183
Pd	6.9	1828
W	18.3	3695

outlined in chapter 5, were CO_2 gas cooled, and one of them (Calder Hall) was the world's first commercial nuclear power plant. However, in a water-cooled reactor this alloy would not meet one of the criteria above, chemical compatibility, as it reacts with water. A secondary, if not equally important, consideration is the limitation in operation to 633 K, reducing the efficiency of the plant. In the next generation of CO_2-cooled reactors, the AGR, the cladding was changed to an Fe-based one.

Since silicon-based cladding would have similar mechanical limitations to carbon, the next choice would be zirconium, which is currently the basis for most cladding today in LWR.

3.7 Zirconium-based cladding

Zirconium meets most of the criteria set out above for use as cladding, with one slight caveat, it does react slowly with water, but the reaction is manageable with control of the water chemistry. The primary advantage of using Zr as cladding is that enrichment of the fuel is not needed, for example the CANDU reactors can use natural uranium as fuel, but in this case they use deuterium to reflect neutrons back into the fuel, increasing fission efficiency [13]. These advantages make the use of Zr-based alloys for cladding prevalent amongst commercial power plants.

3.7.1 Material properties of zirconium

Zirconium, as any metallurgist will tell you, is hexagonal, with a melting point of 2128 K, and it undergoes a phase transition from hexagonal to body-centred cubic at ~1136 K (figure 3.15).

Since Zr is a metal, it has good thermal conductivity, significantly higher than UO_2, for example, across a wide temperature range (figure 3.16). This, therefore, should not cause the Zr to become a heat-limiting zone within the core, reducing reactor efficiency [5].

$$\lambda = 7.51 + (2.09 \times 10^{-2})T - (1.45 \times 10^{-5})T^2 + (7.67 \times 10^{-9})T^3 \qquad (3.2)$$

As outlined above, Zr is hexagonal below 1136 K and consequently there is anisotropy in its thermal expansion. This anisotropy can be used advantageously, in the case of Zr cladding, the material is fabricated such that the *c*-axis preferentially aligns in the thickness, whilst the *a*, *b*-axes align parallel to the tube, shown in figure 3.17. This provides increased tube strength and helps to maximise performance. As a result of the hexagonal lattice, the thermal expansion is now not isotropic, as it is in UO_2, at temperatures below 1136 K, shown in figure 3.18, and this has implications during reactor operation, which we will discuss in the next chapter.

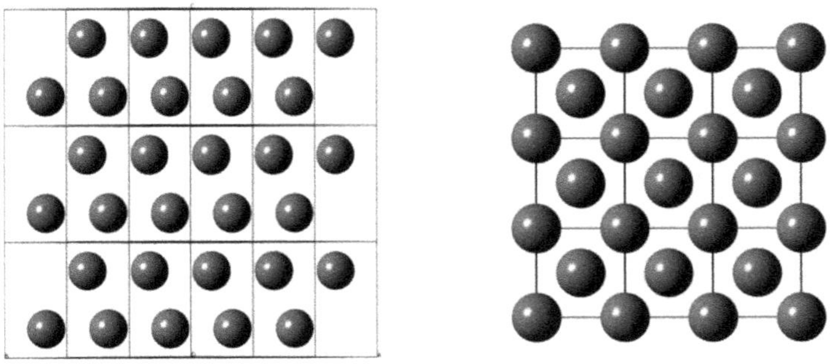

Figure 3.15. Crystal structure of zirconium: (a) hexagonal and (b) body-centred cubic [14].

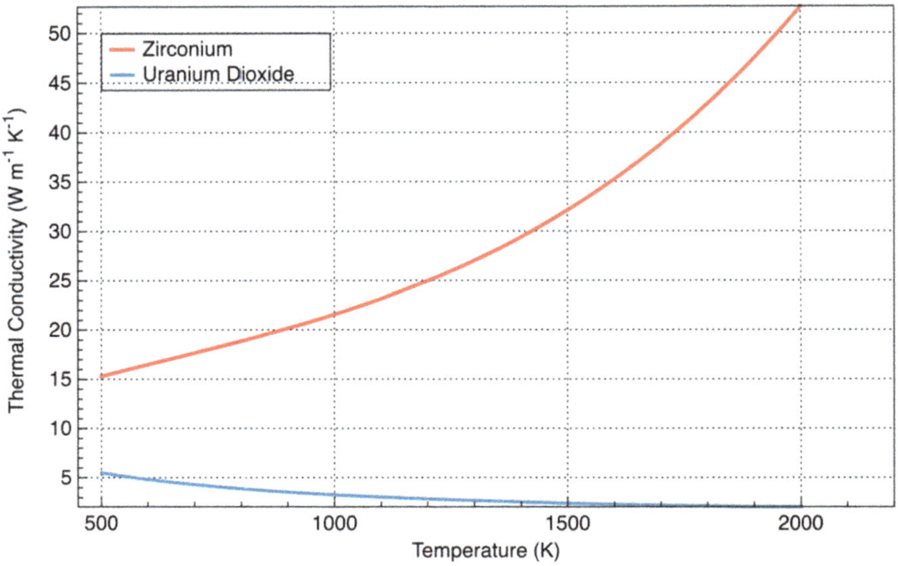

Figure 3.16. Comparison of thermal conductivity for zirconium and uranium.

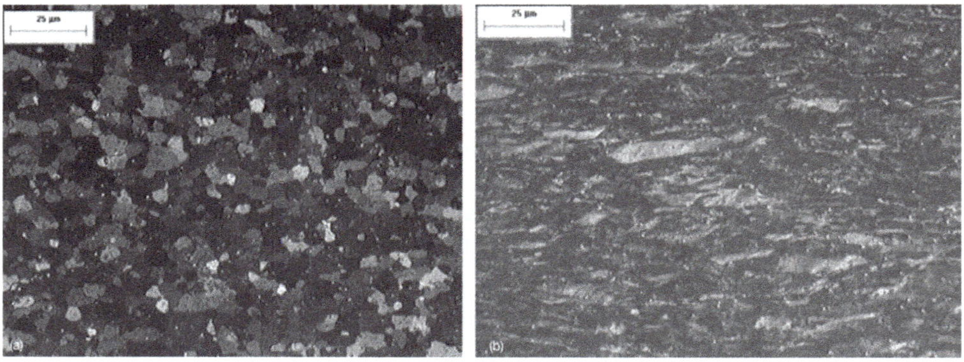

Figure 3.17. The effects from rolling of Zr, highlighting the texture: (a) is recrystallised material and (b) is rolled along the (001) direction [15].

The key drawbacks to using zirconium as cladding relate to the improvement of its chemical stability and mechanical properties in oxidising water-based conditions. These are primarily corrosion based and have resulted in the modification of the composition with added alloying components.

3.7.2 Development of zirconium-based alloys for nuclear applications

During the early days of reactor operation using Zr cladding, it was noticed that there seemed to be a corrosion layer, oxide based, forming on the outside of the tubing. As this was discovered at Chalk River national laboratory, it was given the ubiquitous name of CRUD (Chalk River unidentified deposit) [16]. This layer forming on the surface reduced the thermal transfer of heat away from the elements,

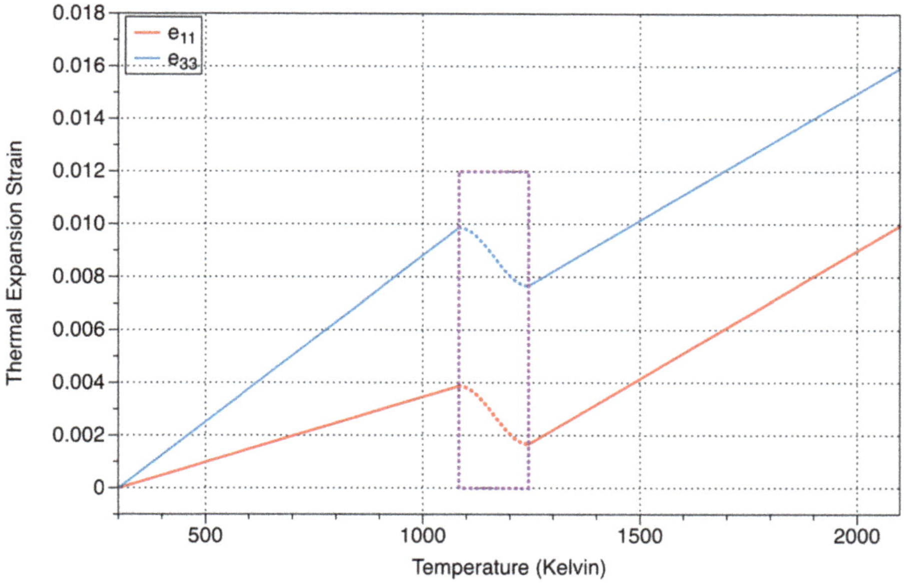

Figure 3.18. The expansion of Zr with temperature in the ε11 and ε33 tensors [5].

Table 3.5. Nominal composition of the five common Zr-alloys. Zirlo is used by Westinghouse, M5 by Areva and E110 by Rosatom. The mass balance in each is Zr.

Alloy	Sn(wt %)	Nb(wt %)	O(wt %)	Fe(wt %)	Cr(wt %)	Ni(wt %)
Zry 2	1.45		0.125	0.14	0.1	0.06
Zry 4	1.45		0.125	0.21	0.1	
Zirlo	1.1	1.1	0.120	0.1		
M5		1.0	0.135	0.038		
E110		1.0	0.06	0.009		

into the water, resulting in trapped heat within the fuel. While the loss was not major, it did present a problem for increased utilisation of fuel.

Analysis of CRUD has subsequently identified the material as being based on (Li, Mn, and Ni)-based ferrite spinels, which have the potential, as we shall see in the next chapter, to impact greatly on the lifetime of cladding in use. The lithium arises from careful management of the water chemistry within a LWR, to minimise corrosion where possible, and also to help manage the neutron flux within the core. As you may recall, both Li and B can be added to absorb neutrons, thus reducing or slowing a reactor's output. As the Zr corrodes, these can be included in the coating formed and reduce a reactor's output even further.

In order to improve the material properties of the clad, different alloying elements have been added, in some cases dependent on the application, giving rise to Zircaloys. The five common ones are shown in table 3.5.

The two common Zr-alloys, Zry2 and Zry4, are designed for different conditions. Zry2 is used in boiling water reactors (BWR), while Zry4 is used in pressurised water reactors (PWR). The degree of corrosion is different in each reactor, as the BWR uses a mix of water/steam, while the PWR, under operating conditions, uses liquid water alone.

During early studies on the corrosion of Zr-alloys, the corrosion products found on the surface were thought to have arisen from impurities in the metal. They was later determined to be from the corrosion of the alloy itself, and this led to changes in reactor operation.

3.7.3 Zirconium hydridation

One the key problems with Zr-alloys is the uptake of hydrogen into the alloy and its modification of properties. Zirconium hydrides form predominantly as ZrH and ZrH_2 intergrowths within the lattice, modifying a range of properties and potentially leading to crack formation in the material. As can be seen from the phase diagram in figure 3.19, Zr can accommodate significant amounts of hydrogen within its lattice at elevated temperatures, but much less at reactor operation temperatures.

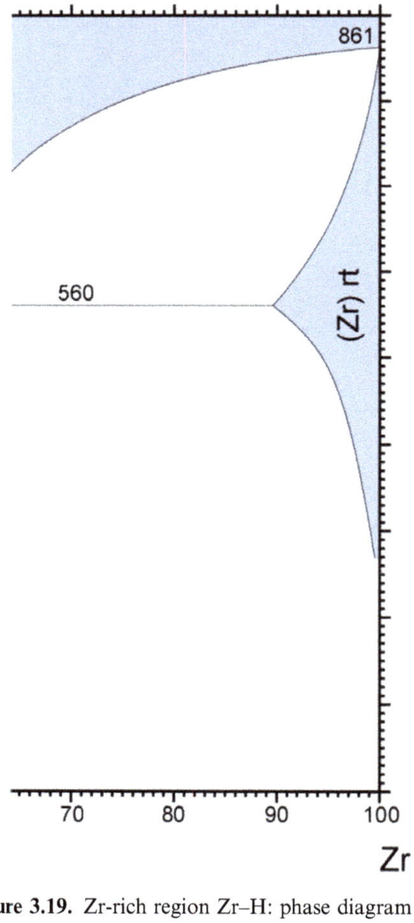

Figure 3.19. Zr-rich region Zr–H: phase diagram [17].

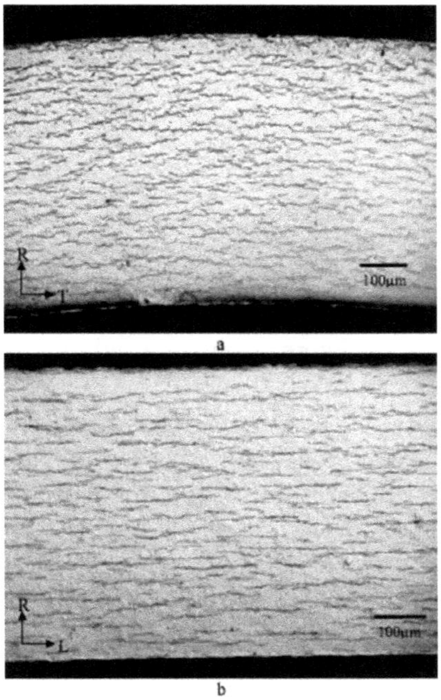

Figure 3.20. Images of Zr–H formation in Zr-alloys [15].

The key drawback of hydride formation within Zr is the difference in structure adopted; both ZrH and ZrH_2 are orthogonal in structure, with ZrH being orthorhombic and ZrH_2 tetragonal. As such, these phases have different properties and are different from the hexagonal Zr lattice. Consequently, there now exist sources of stress and strain within the lattice, reducing the strength and embrittling the material, and often leading to crack formation and propagation, examples are shown in figure 3.20.

There is a further implication that can arise from Zr–H formation, and we will discuss this later when discussing cladding under operation within the core.

One advantage of using Zr as the basis for cladding is the low neutron absorption, but if you make an overall economic assessment of the reactor in operation, an alternative approach can be used. What if we take a loss on the neutron efficiency, but save money on the operational costs of the clad? Zirconium is not a cheap element and is often found with hafnium, and since Hf has a high absorption of 100 barns (Zr is 0.185) [1], it needs to be removed before use. Hafnium is often used as a neutron poison in waste applications preventing criticality, hence it is not ideal for use within a core.

If we accept enrichment of fuel, but can save money in other areas, other elements become usable. If you also factor in that in gas-cooled reactors, designed to operate at temperatures above 900 K—above the hcp–bcc transition in Zr, making Zr non-usable—these other elements become viable. One such element is iron, which has an absorption of 2.56 barns, but which has other advantages.

3.8 Iron-based cladding

Iron, while having a higher absorption than Zr, has many advantages when used within reactors. In one particular case, it is highly advantageous when compared with zirconium: most Fe-based alloys are cubic, and therefore will have isotropic thermal expansion. Other advantages include:

(i) increased strength when compared with Zr, it is therefore capable of being used at a thinner thickness, accommodating the increase in neutron absorption;
(ii) an extensive ability to be machined and handled;
(iii) a history of alloying, with many elements able to match the design requirements, for example, stainless steel with increased corrosion resistance.

Each of these advantages helps to offset the reduced ability for neutron transport through the clad, although, in many cases, Fe-based alloys are not used in LWR, and the major use has been within the AGR in the UK, which use enriched fuel. In these reactors, which use CO_2 as the coolant/moderator, the steel-based clad has unique advantages.

3.8.1 Advanced gas-cooled reactor cladding

These reactors require that certain criteria be met by the cladding when in use, two of which are: the ability to operate at high temperature and the ability to show minimal corrosion/reduction by CO_2 gas. As such, the cladding is a non-standard steel composition, as shown in table 3.6.

As can be seen in table 3.6, the additives to the alloy have similar neutron absorptions to Fe itself, and so do not greatly impact on the neutron transparency. Given the nature of the environment within an AGR, the cladding chosen, i.e. 20/25-stainless steel, optimised multiple requirements, but did require enrichment of the fuel to ~3.5% before use. This compares with the enrichment in the Westinghouse AP1000 of 4.95% ^{235}U [19], however, this is for economic reasons.

3.9 How do fuel and cladding relate to each other?

Generally, fuel is stacked within cladding to form fuel rods, the length of which is determined by the reactor design, for example in the AP1000 design the length is

Table 3.6. A compositional analysis of the cladding used in AGR. The mass balance is Fe [18].

Cr (wt%)	Ni (wt%)	C (wt%)	Mn (wt%)	Si (wt%)	Nb (wt%)	S (wt%)	Ti (ppm)
20.4	25.1	0.02	0.7	0.56	0.60	0.007	<130

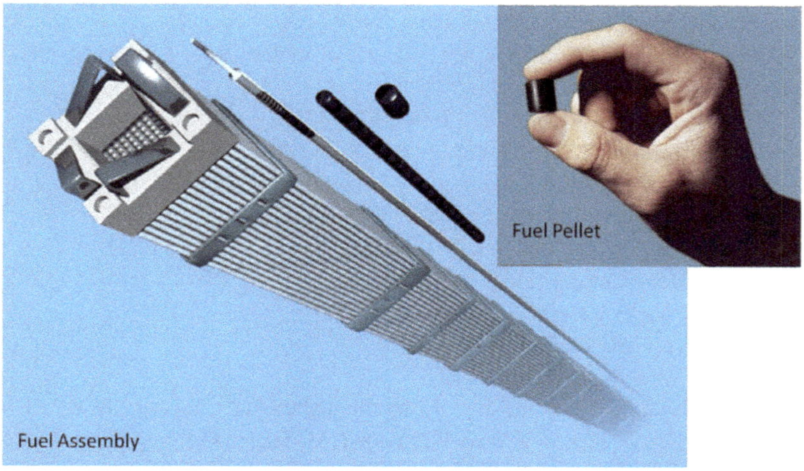

Figure 3.21. Schematic showing the fuel assembly stages, from pellets through to completed assembly [22].

~3.65m [19]. These rods are then combined with others to form fuel bundles, and again the exact nature of the bundle depends on the reactor design. These bundles themselves are then combined into fuel assemblies, which are then placed into the core. The majority of fuel cores are based on a circular approach, with an array of bundles, ranging from 10 × 10 in an advanced boiling water reactor (ABWR) [20] through to 17 × 17 in an AP1000 [19]. A schematic is shown in figure 3.21.

When placed into a reactor, fresh fuel is generally placed away from the centre of the core, and is then periodically moved closer to the centre. One of the early patents for a reactor core design proposed this, with the oldest fuel being placed in the centre of the core with less used fuel around it [21]. Having old fuel in the centre of the core, with fresh fuel further away, helps to balance the reactor's output, reducing the likelihood of hotspots in fission. Each reactor handles its fuel differently, but it is often the case that a cycle between fuel replacement is about 18 months. For example, in the Westinghouse AP1000, the core contains 157 fuel assemblies, with an average change of ~68 every 18 months [19].

In some reactor designs, for example the AGR, there is the potential for online refuelling without the need to shut down the reactor.

So, in this chapter we have discussed the requirements for fuel and cladding, and how they are chosen and designed for use. In the next chapter we shall discuss the effects on material properties within the core.

References

[1] Chadwick M B *et al* 2006 ENDF/B-VII.0: next generation evaluated nuclear data library for nuclear science and technology *Nuclear Data Sheets* **107** 2931–3060
Chadwick M B *et al* 2011 ENDF/B-VII.1 nuclear data for science and technology: cross sections, covariances, fission product yields and decay data *Nuclear Data Sheets* **112** 2887–996

[2] Chiotti C P, Klepfer H H and White R W *Trans. Am. Soc. Metals* **51** 772–86

[3] Chevalier P Y, Fischer E and Cheynet B *CALPHAD* **28** 15–40

[4] Martin D G *J. Nucl. Mater.* **152** 94–101
[5] IAEA 2006 Thermophysical properties database of materials for light water reactors *IAEA TECDOC* 1496
[6] Willis B T M *J. Chem. Soc. Faraday Trans.* **83** 1073–9
[7] White J T and Nelson A T *J. Nucl. Mater.* **443** 342–50
[8] Guéneau C, Chatillon C and Sundman B *J. Nucl. Mater.* **378** 257–72
[9] Adamson M G, Aitken E A and Caputi R W *J. Nucl. Mater.* **130** 245
[10] Gibby R L *J. Nucl. Mater.* **38** 163–77
[11] Nuclear Regulatory Commission
[12] Higgins G T *J. Nucl. Mater.* **8** 153–9
[13] Olander D *J. Nucl. Mater.* **389** 1–22
[14] Russel R B *J. Nucl. Mater.* **6** 1045–52
[15] IAEA 2010 Delayed hydride cracking of zirconium alloy fuel cladding *IAEA TECDOC* 1649
[16] Short M P, Hussey D, Kendrick B K, Besmann T M, Stanek C R and Yip S *J. Nucl. Mater.* **443** 579–87
[17] Vaughan D A and Bridge J R *J. Metals* **8** 528–31
[18] Ecob R C, Lobb R C and Kohler V L *J. Materials Science* **22** 2867–80
[19] Nuclear Directorate *Generic Design Assessment—EDF and Areva—Step 3 Fuel Design Assessment Report*
[20] Nuclear Directorate *Generic Design Assessment—Step 2 Assessment of the Fuel and Core Aspects of Hitachi-GE's UK Advanced Boiling Water Reactor (UK ABWR)*
[21] Stern F M 1969 Core design for nuclear reactor *US Patent Specification* 3432389
[22] Duke Energy

Chapter 4

Nuclear fuel, part 2: operational effects

As with chapter 3, the key requirement in any nuclear reactor is fuel. How the fuel, and cladding, if used, behaves within a core is one of the key criteria when running a reactor. The long-term effects can not only be detrimental, but can also ultimately limit the efficiency of a reactor. The effects of fission on fuel have led to new options being developed, which while they address some problems very effectively, give rise to other issues that need to be addressed.

Before we consider the effects of fission, we should actually define the key stages that occur in nuclear fuel during the initial stages of burn up.

4.1 Initial stages

As any solid state scientist—whether they be a chemist, physicist or materials scientist—knows, as you heat ceramics to high temperatures, common processes occur. These are thermal expansion, followed by, if the temperature is high enough, sintering (densification) and grain growth. When fuel pellets of UO_2 are initially formed, they are sintered to ~2200 K to ensure that they meet the density required for use within a fission core. Under normal conditions, this temperature should be high enough to prevent any further sintering or grain growth, but as has been outlined before, a fission core is anything but normal. If one considers the centre temperature of a fuel pellet (~1500 K), it is sufficiently high to allow grain growth and further densification to occur, although this would not be rapid. However, once the temperature increase from fission is included in the process, we reach conditions that allow further densification and grain growth to occur. These processes can cause problems for the long-term use of fuel, but why?

Consider a fuel pellet purely in terms of thermal conduction and output: larger grains could be beneficial, providing more rapid conduction of heat away. However, we are not just considering thermal conduction, we also need to take

doi:10.1088/978-0-7503-1104-5ch4

into account a range of other factors. These factors can be broken down into two main areas:

(i) 'classical'—the effects that would occur in normal heating of ceramics to high temperatures, and
(ii) 'fission'—these are the effects that arise from the fission process itself and the products formed.

Before we discuss the consequences of nuclear fission and fission product formation, we should first explore the classical effects expected from heating ceramic pellets to high temperatures.

4.2 Classical effects from heating

As outlined above, two of the key effects when heating ceramics to high temperatures are sintering (or densification) and grain growth. In many cases, they relate closely to each other, with grain growth occurring from sintering, the joining of grains together, but they can occur separately. The key point to remember is that the nature of ceramics can change, and in the case of nuclear fuel, move away from the designed properties.

Generally, ceramics with an average small grain size will shrink when heated, as the grains sinter together and form larger grains. These grains can then merge with other grains, or, as is often the case, absorb smaller grains, schematically shown in figure 4.1, with an example from UO_2 in figure 4.2. An animation of the process is show in figure 4.3.

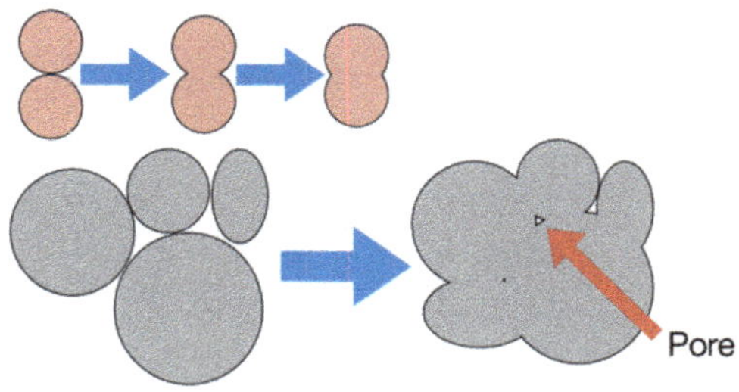

Figure 4.1. Schematic of the sintering process: (a) two single grains and (b) multiple grains.

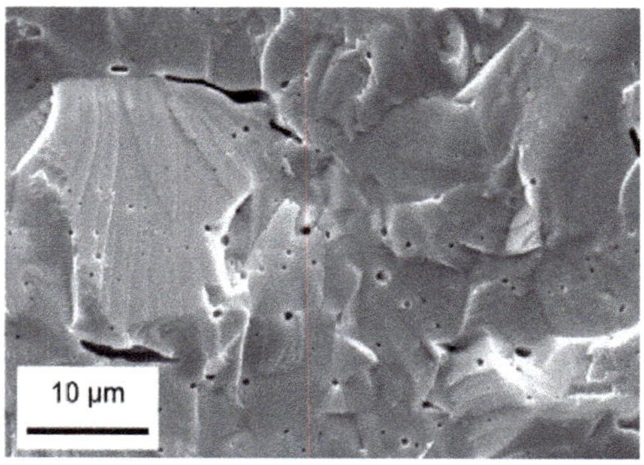

Figure 4.2. UO_2 after sintering and before irradiation [1].

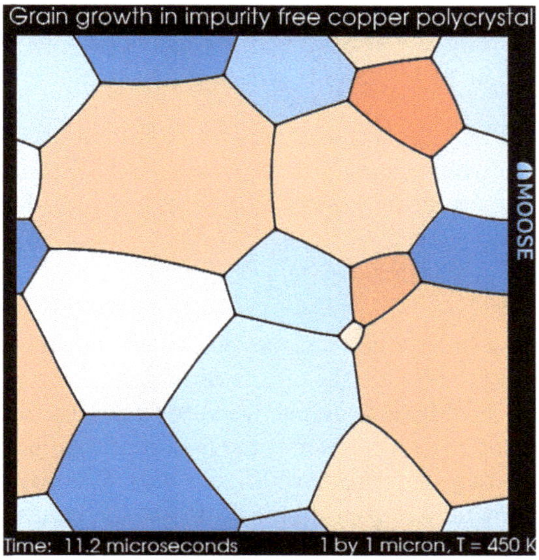

Figure 4.3. INL Moose animation on grain growth. Courtesy of Michael Tonks.

4.2.1 Why does sintering occur?

The answer to this question is based around both thermodynamics and chemical reaction. If there are a large number of small grains within a system, each of these grains will have a surface, and each of these surfaces will have an energy associated with it. If the total surface area within a system is reduced, then thermodynamically the system should be more stable, as it has a lower surface energy. In addition, sintering may occur, as there is a further chemical reaction occurring at the interface, and this is how many oxides can be prepared on the industrial scale.

4.2.2 What are the outcomes of sintering?

Thermodynamically speaking, the perfect sintered lattice is a 100% dense material, with no pores, containing multiple grains with different orientations. Each of these grains would have an overall grain orientation of zero, i.e. there would be no polarisation in the structure. However, perfect lattices tend not to arise, and with many cases defects are found; these defects can be atomic scale, e.g. vacancies/interstitials, or mesoscale, e.g. voids. In addition to these effects, the samples generally shrink, with the shrinkage depending on multiple factors, not least time at temperature and initial density.

One factor with pellets that are heated is the release of stress/strain within, this can manifest itself in many ways, but the methods that concern us most are cracking and shape change. As outlined when discussing radiation damage, one method for a system to accommodate a point of stress/strain and relieve the associated energy, is to crack. It is the nature of the crack that concerns us most with regards to fuel. The second method is to induce a shape change, changing from a pellet with straight sides to one that closely resembles an 'hourglass' (figure 4.4). This reduces the plane strain in the middle of the pellet, and the plane stress at the top and bottom. Generally, this is associated with cracking within the pellet.

Do we want a 100% dense material as a fuel matrix?

4.2.3 Why do we care?

The key point in answering the question of why this is important, is that we need to minimise as much as possible changes in the base criteria within the fuel. This is key, as in the next section, once the effects of fission are included, fuel can undergo substantial changes.

4.3 Fission products

So far in this chapter we have concerned ourselves with the classical effects of heat on solids and we now need to include the effects of fission on the fuel. However, before we can discuss the effects of fission on fuel, we need to discuss the nature of the fission products formed. As outlined in chapter 1, the fission process involves the splitting of an atomic nuclei into two or more parts (figure 4.5). We now need to discuss the nature of these parts, as they impact dramatically on the behaviour of fuel.

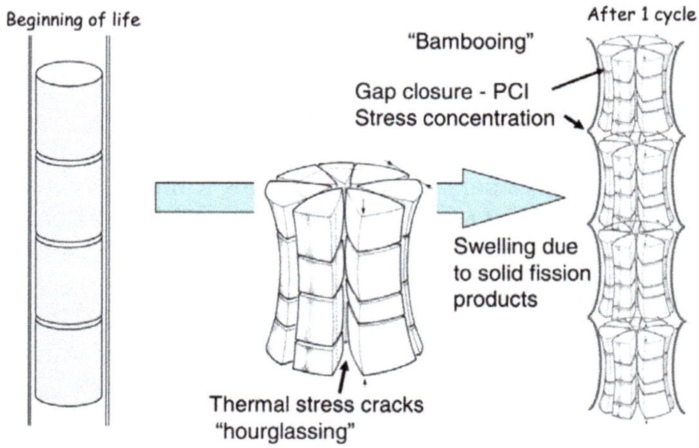

Figure 4.4. Hourglass formation within a fuel pellet [2].

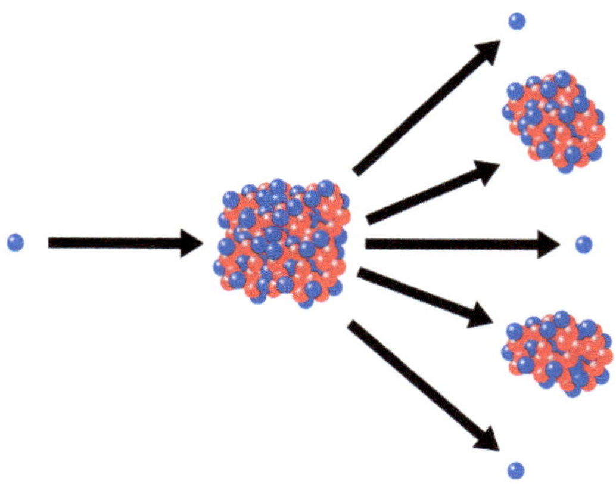

Figure 4.5. Schematic of the fission process.

Fission is a quantum mechanical process and if we assume a probabilistic approach to predicting the outcomes, we can obtain a probability of fission yield for an isotope. An example is shown in figure 4.6, where the fission yield probability for both ^{235}U and ^{239}Pu is plotted.

As can be seen in figure 4.5, the probability has two maxima centred around 90–100 and 130–140 amu. Whilst each fission event is unpredictable in terms of what is produced, if a large enough volume of material is examined, an average probability of the fission products can be predicted.

As can be seen in table 4.1, there are some likely predicted elements/isotopes that will form during fission. It is the nature of these products that helps to influence the long-term nature of fuel. These products vary in nature, some are stable and can

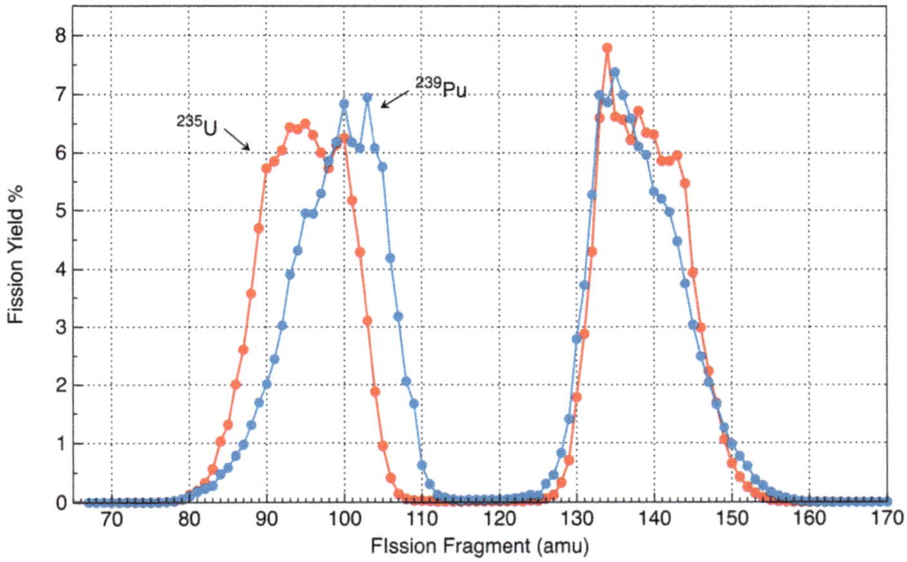

Figure 4.6. Probability of fission product production for ^{235}U and ^{239}Pu. Data taken from [3].

Table 4.1. Selected fission yield probabilities predicted for ^{235}U [3].

Fragment(amu)	Isotope	Half-life	Absorption(barns)	Yield (%)
133/134	Cs	Stable/2.065 yr	29/140	6.8
135	I/Xe	6.57 h	2650000	6.3
93	Zr	1.53 my	2.7	6.3
99	Mo/Tc	66 h	20	6.1
137	Cs	30 yr	0.11	6.1
90	Sr	29 yr	0.90	5.75
131	I	8 d	80	2.8
149	Sm	Stable	15200	1.1
85	Kr	11 yr	1.66	0.3
155	Eu/Gd	4.76 yr/stable	3950, 60900	0.03

Table 4.2. General classes and examples of fission products.

Class	Elemental examples
Dissolved	Ba, Zr, Y and lanthanides
Precipitated	Ba, Sr, Sn, Zr, Mo(Tc), Cs and Rb
Metal particulates	Pd, Ru, Rh, Pt and Mo(Tc)
Volatiles/gaseous	He, Kr, Xe, Te, I, Br and Cs

cause problems for operation, the classical one being $^{139}I/^{139}Xe$, while others are transient and decay, for example ^{131}I, and there are even a few that have medical uses, for example, $^{99}Mo/^{99m}Tc$.

In general terms, fission products can be broken down into four main classes, as shown in table 4.2.

As can be seen in table 4.2, the nature of some the elements can vary, for example Ba and Cs, and this is primarily due to the nature and composition of the fuel. For example, the solubility of Ba within UO_2 increases with temperature, and so it may dissolve within the matrix, but at lower temperatures it begins to exsolve out of solution, forming precipitates. The key thing is to remember that each fission product can modify the properties of the fuel.

The two classes of fission product that can cause some of the biggest problems are metal particulates and volatiles/gaseous, each of these can form on their own and have implications, but they are often found together, with an increased impact on fuel. Before we discuss their combined impact, let us consider the issues arising from their formation individually.

4.3.1 Volatiles/gaseous

As the name suggests, these are elements that are either highly volatile or gaseous under the conditions within the reactor. As such, they can be extremely mobile within the lattice, and can agglomerate together, for example forming gas bubbles of He, Kr and Xe. Such bubble formation can induce stress/strain within lattice, often, but not necessarily, forming at pre-existing locations of stress/strain. As we discussed in chapter 2 during the discussion on recovery from damage, such bubble formation can have a dramatic impact on the lattice.

4.3.2 Metal particulates

These, as shown in table 4.2, tend to be the noble metals, which are known to have low reactivity in many situations. Their primary impact is the modification of material properties through the introduction of stress/strain within the matrix.

4.3.3 Combined effects

It is often found within fuel that gas bubbles tend to form in co-existence with metal particulates in the lattice, as shown in figure 4.7. These regions can have a dramatic impact on the stability of the fuel, as, if we think back to the discussion on radiation

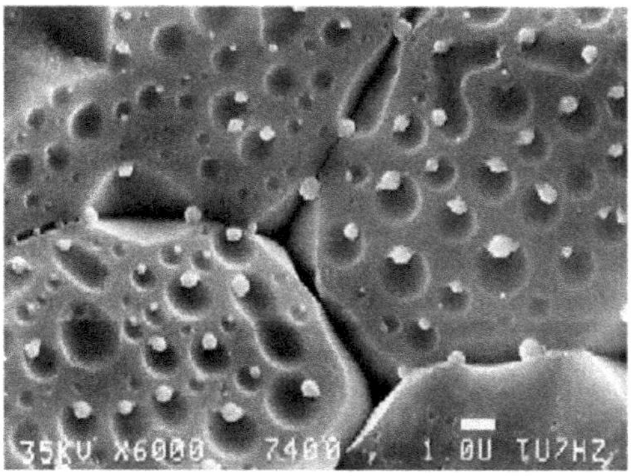

Figure 4.7. Image showing metal particulates with co-located gas bubbles [4].

damage and crack formation in chapter 2, we are now providing more possibilities for crack initiation in the material.

So, having introduced the nature of fission products and discussed how they can impact on the lattice, we can now turn our attention back to the effect of fission on fuel.

4.3.4 Key impacts of fission

When considering nuclear fission, one key impact springs to mind—where there was once one atom, there are now at least two. This new atom causes problems for the material, irrespective of the atom effects, such as temperature rise and damage from translating particles (fission products, neutrons, etc). Now that we understand the effects of temperature and fission on ceramic fuel as separate issues, what happens if we combine the two together?

4.4 Initial reactor operation

In most cases, reactors are fuelled at room temperature, and after they have been refueled, the control rods will be withdrawn and the fission process will be induced and allowed to run. The process is outlined in figure 4.8.

As can be seen in figure 4.8, during initial startup the fuel will undergo densification and fission products will be formed; one process will shrink the pellet, while the other wants to expand it. These in themselves will introduce cracking, which will affect the material properties of the fuel. One issue we have not fully discussed with regards to reactor operation, is the effect on thermal conductivity of these effects; put simply, the thermal conductivity reduces, potentially leading to trapped heat within the fuel and a reduction in the efficiency of the heat transfer to the coolant.

So far in this discussion we have not included the effects of radiation damage, which in itself has the potential to dramatically alter the structural properties of the fuel. For example, as we showed in chapter 2, damage not only effects crystallinity, but it can also induce creep, induce stress/strain with the lattice, and reduce the thermal conductivity of the material, with the ultimate possibility of amorphisation.

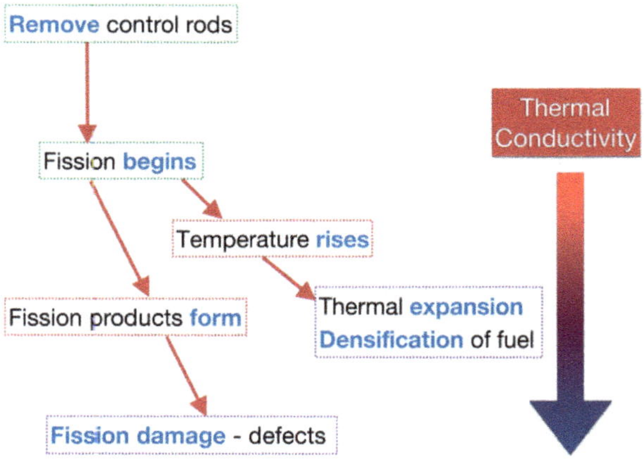

Figure 4.8. Schematic of the initial stages in reactor operation.

When comparing levels of damage within nuclear fuel, the classical unit of displacement per atom (dpa), while still having value, becomes somewhat tricky to estimate. A simple comparison is this: while we could use dpa, fuel within a core can experience a minimum of 1 dpa per day, so in the process of operation within the core, given an average lifetime in the core of six years, we have ~2200 dpa. A reactor does not have a single energy spectrum within the core, and with powering up/down in reactor cycles, it is not easy to determine. A different unit, which is based on reactor output, is the GWDthm, which uses the power output of the fuel within the core as a metric for comparison.

As discussed above, one of the key fission products is volatile and can lead to the formation of gas bubbles. These gas bubbles can act as sources for crack initiation and reduce the thermal conductivity, leading to changes in the microstructure of the fuel, an example of which is shown in figure 4.9, and enhanced hourglassing of it, as shown in figure 4.4.

This changed microstructure, known as high burnup structure (HBS) or cauliflower fuel (figure 4.9(c)), shows multiple changes from fresh unused fuel, closely following the properties expected from damage:

(i) decreased hardness;
(ii) decreased thermal conductivity;
(iii) increased porosity;
(iv) expanded lattice parameters.

Each of these modifications would cause problems in themselves in the long-term use of fuel, but combined the effects are magnified. If we consider the effects on fuel of use within a core, we can obtain a relationship similar to that found in figure 4.10.

If we focus on the effects on thermal conductivity, then as the operational lifetime of the fuel increases, i.e. is in the reactor longer, the thermal conductivity decreases dramatically (figure 4.11).

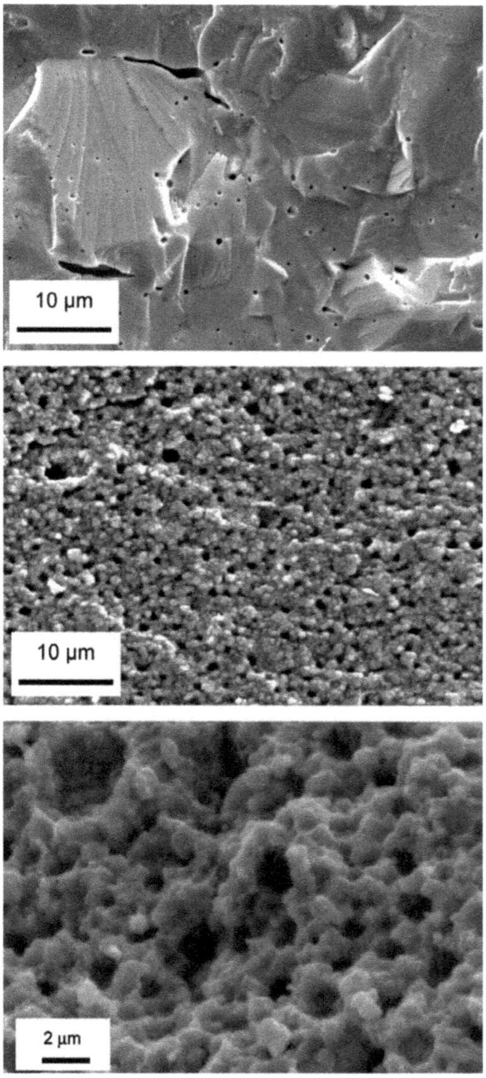

Figure 4.9. The change in microstructure of UO_2 fuel at (a) 0 GWD thm, (b) 75 GWD thm and (c) 160 GWD thm [1].

One question that was posed earlier and that we have not answered is: *do we want a 100% dense material as a fuel matrix?*

The answer to this question can be complicated or simple to answer, but the answer can be concentrated down to: *we want the fuel to be as dense as necessary for extended reactor operation*, which means that while a dense fuel is better for thermal conductivity, we want the material to be able to withstand the effects of operation within a reactor for as long as possible. If the material were 100% dense, with increased fission and product formation, there would not be voids where fission products could locate. In a material that has pores/voids, these can act as nucleation

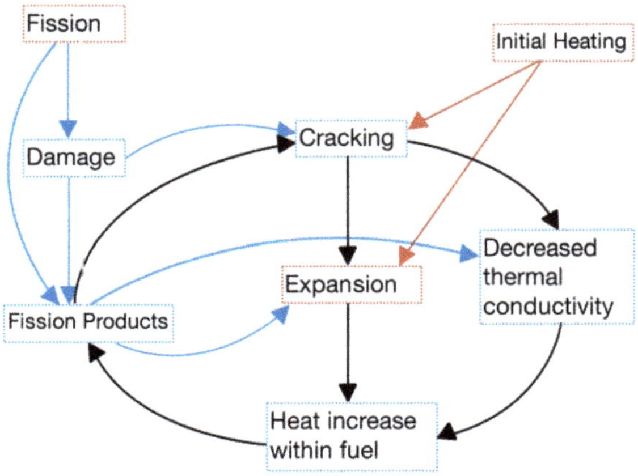

Figure 4.10. Schematic of fuel interactions within a reactor core.

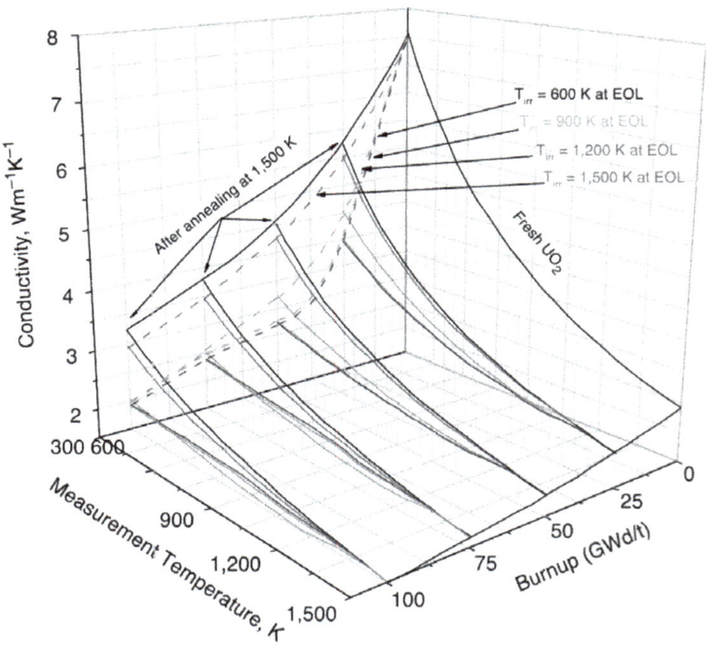

Figure 4.11. The effect on thermal conductivity with increasing burn up of UO_2 [5].

sites for gas atoms to agglomerate, reducing, if only to a somewhat minor extent, the expansion that would occur.

So far we have only discussed the effects of operation on the fuel pellets themselves, neglecting the cladding. We should now include the effects of operation in the core on the cladding, and look at how the cladding interacts with the fuel.

4.5 Fuel cladding under operation within the core

In chapter 3 we considered the common types of cladding, predominantly Zr based, and for the purposes of this discussion we will consider only Zr-based alloys.

If you remember, one of the advantages of using Zr-based cladding is the low neutron absorption, however, this does not mean that the cladding will not interact with neutrons—it does. The effects from the damage are different from those experienced within the fuel pellet, but they still affect the fuel indirectly. If you remember from chapter 2, some of the key effects of damage are the potential formation of defect loops within metals, expansion and the other bulk properties. In order to understand the effects of these within clad within a fission core, we shall discuss them separately and then combine them with the changes observed in fuel.

4.5.1 Damage formation and expansion

As you may recall from chapter 3, Zr-sheet is generally rolled such that the material is highly textured in the *a,b* plane. This affects the predominant direction for radiation damage in clad, where, as in a cubic system, the likelihood of damage will effectively be isotropic; in Zr, which is hexagonal, there is a propensity for damage to occur in the *c* plane. This damage tends to give rise to the formation of defect loops in the *c* plane, i.e. parallel to the surface of the clad.

One of the effects of damage is a volume increase, or expansion, and in the case of Zr this can be predicted using an empirical equation (equation 4.1) [6]. This relates the expansion to the neutron flux, the texturing of the sample, the temperature and the reduced cross-section area.

$$\frac{\Delta L}{L} = a\left[\exp\left(\frac{240.8}{T}\right)\right](\phi t)^{\frac{1}{2}}(1 - 3f_z)(1 + 2\mathrm{CW}). \tag{4.1}$$

Luckily, the degree of expansion is small, but it can be still be problematic.

4.5.2 Hardness and ductility

As outlined previously, in general the hardness increases with damage, while the ductility decreases. These are both detrimental to the long-term use of materials, however, in these materials they generally saturate, i.e. do not increase, at neutron fluencies of $\sim 10^{24}$ n m^{-2} [6].

Since the Zr cladding undergoes changes to its properties with increasing levels of damage or time in the core, how do these compare to those of the fuel pellets themselves?

4.6 Fuel and cladding

As was outlined in chapter 3, fuel is stacked within cladding, generally with a small gap between the two. If we now consider the initial stages of reactor operation (figure 4.8), it is possible to understand why some of the design decisions were made. During the initial heating stages, both the cladding and the fuel expand, as shown in figure 4.12, with different rates of expansion with temperature, giving rise to different levels of expansion. The gap between the fuel and cladding allows for the different levels of expansion and provides a means to stop the fuel and cladding from interacting. Is this always the case, though?

If you remember back to earlier in the chapter where we discussed densification and hourglass formation, we outlined that while the pellet may shrink in volume due

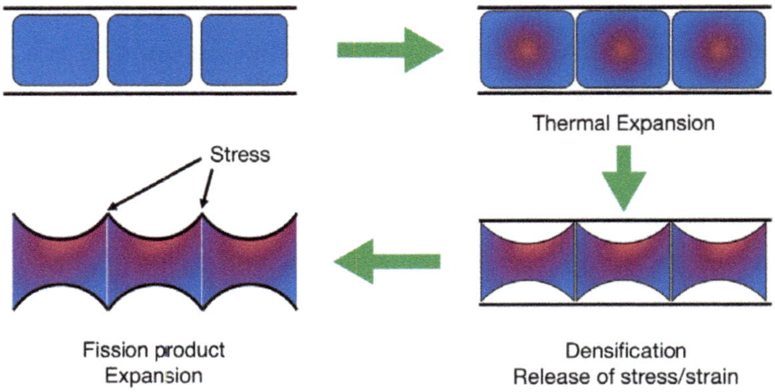

Figure 4.12. Schematic of fuel bamboo formation.

to densification or sintering, it may crack and change shape. This new shape is similar to an hourglass, hence the name. This process should therefore imply that with temperature the pellet could change shape. Now also cast your mind back to where we discussed the effects of fission product formation on a pellet, with the associated increase in volume. How would these processes interact, within the pellet and with the cladding? The simple answer would be, as shown in figure 4.12 and outlined below, not ideally.

During use fuel and cladding go through the following stages:

(i) both the fuel and clad will thermally expand—there may be interaction between the pellet and the cladding;
(ii) the fuel pellets will begin to sinter/densify and reduce stress/strain—potentially forming an hourglass shape;
(iii) with increasing formation of fission products, the fuel pellet will increase in volume and interact with the fuel cladding—giving rise to 'fuel bamboo'.

From this process, we now have two further complications that can be introduced, arising from the interaction between the fuel and the clad:

(i) pellet–clad interaction (PCI), and
(ii) stress corrosion cracking (SCC).

4.6.1 Pellet–clad interaction (PCI)

This is, as the name suggests, an interaction between the fuel pellet and the cladding. It arises from the mismatch in material properties, which, as was shown previously, is a cause of fuel 'bambooing'. Now that the fuel is in contact with the cladding, there is the potential for reaction between the mobile fission products and the cladding, further weakening the cladding integrity.

Before we discuss the second of these two further complications, stress corrosion cracking, we should understand the interactions of cladding with water.

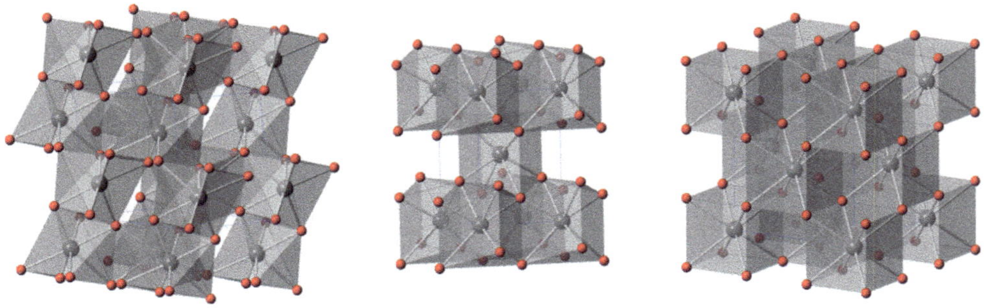

Figure 4.13. Crystal structures of ZrO_2 in (a) monoclinic, (b) tetragonal and (c) cubic forms [7].

4.7 Cladding corrosion

When it comes to Zr-based cladding, one of the key interactions, if not the most important one, is between the clad and the coolant. Generally, the coolant is H_2O- or D_2O-based and can lead to some interesting interactions between the cladding and the coolant.

As was outlined in chapter 3, zirconium can pick up hydrogen-forming hydrides, and what can also occur is oxygen dissolution into the lattice, forming nano-regions of zirconium oxides (ZrO_2). However, the biggest concern is the reaction Zr can have with H_2O, which itself can have multiple impacts. It should be noted, however, that while Zr is classically considered to be a non-reactive metal, at the elevated temperatures within a reactor core, it can and does oxidise.

The reaction between Zr and H_2O can be thought of as a simple displacement reaction, with the products being ZrH_x, ZrO_2 and H_2:

$$Zr_{(s)} + H_2O_{(l)} \rightarrow ZrH_{x(s)} + ZrO_{2(s)} + H_{2(g)}. \tag{4.2}$$

Each of the three products outlined above has implications for the long-term stability of Zr-based alloys. We have already discussed the implications of ZrH_x inclusion within Zr, and how that modifies the properties. The formation of the oxide layer has both advantageous and disadvantageous effects on the matrix.

One the advantages to forming ZrO_2 on the surface of Zr is that, as with most oxide coatings, it reduces the rate of corrosion in the base material, in this case Zr. Therefore, it would be expected that as ZrO_2 forms, the rate of reaction will slow, and this does indeed occur. In the case of ZrO_2, there are, as you can probably expect, complications arising both from the nature of the crystal structure and thermal conductivity. For the sake of this discussion, let us deal with the thermal conductivity first.

As outlined in chapter 3, the thermal conductivity of ceramics is generally lower than that found in metallic systems; see the discussion about the conductivity of Zr and UO_2. Therefore, using the same methodology for ZrO_2, we can see that the thermal conductivity will be lower, which reduces the heat transfer out of the cladding. One application arising from this effect is the use of ZrO_2 for thermal barrier coatings in exhausts.

Zirconium oxide is both simple and complex due to its ability to adopt one of the three crystal structures shown in figure 4.13. At low temperatures it adopts a monoclinic crystal structure, based on the mineral baddeleyite, which transforms to a tetragonal lattice at ~1450 K, with a further transformation to a cubic lattice, identical to that adopted by UO_2, fluorite, at ~2650 K. A simple approach would be to expect that ZrO_2 will form in the monoclinic phase, however, this is not always the case, and often it can form in both the cubic and tetragonal forms. They form for a variety of reasons, the two most relevant here are:

(i) the surface energy of the system—as the grains are small, the surface energy provides stability to what are thermodynamically unstable phases, and therefore considered meta-phases;
(ii) the strain introduced by the Zr matrix—as with surface energy, this now provides the extra energy to further stabilise the tet/cubic forms.

As the film begins to form on the surface, these phases become increasingly unstable, as the grain size is increasing, thus reducing the surface energy stabilising, and the extra stability from the matrix is becoming proportionally smaller. The outcome of these phases becoming unstable is that they begin to spall off, i.e. they form small particles that break away from the oxide layer.

This spalling now can have the effect of reducing the passivation on the surface of the Zr-matrix, which begins to increase corrosion again.

4.7.1 Corrosion mitigation and CRUD

It is possible to modify the chemical behaviour of the cooling water, for example boron and lithium are already added to act as a moderator (B) and reduce acidity (Li). These can help to slow the corrosion process, and in many designs the corrosion is easily manageable. In the early days of reactor design and use, it was noticed at the Chalk River Nuclear laboratories in Canada that a coating was forming on the cladding they were using and examples are shown in figures 4.14 and 4.15. This gave rise to the acronym CRUD (Chalk River unidentified deposit [8]) and was thought to arise from impurities within the alloys. It has since been identified as Ni,Fe-based spinels [8] and it can lead to similar problems to those outlined above for ZrO_2.

Figure 4.14. Image of corrosion on the surface of cladding within a LWR [9].

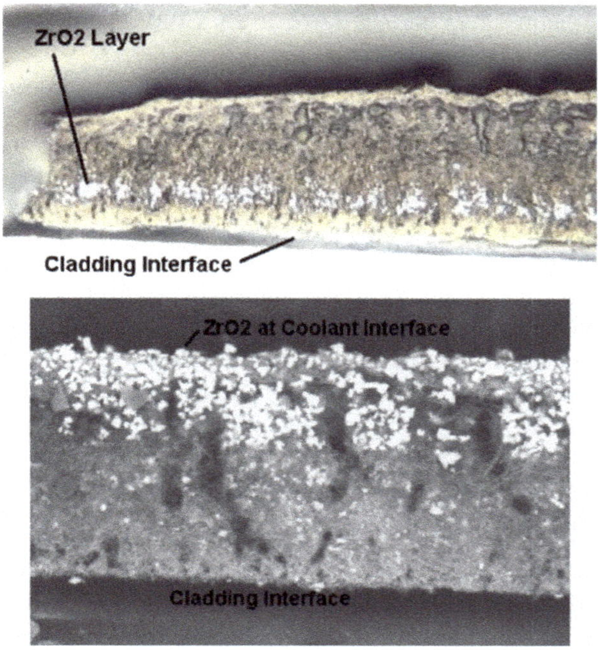

Figure 4.15. Light micrograph and SEM image of CRUD—the ZrO_2 can be seen in both images, highlighting the oxidation of the base cladding [9].

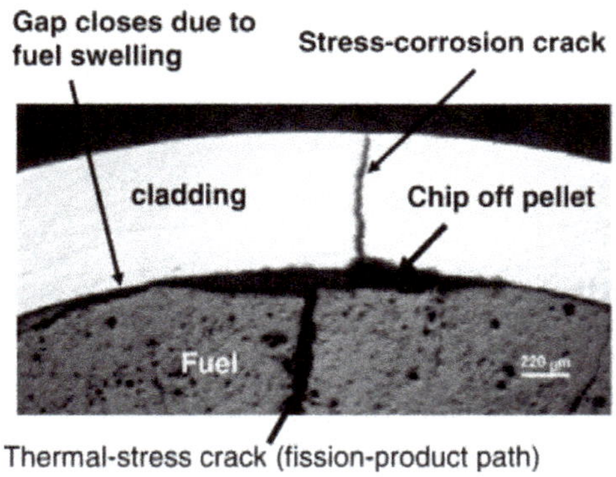

Figure 4.16. Image of a crack induced in cladding corresponding with a crack in a fuel pellet [2].

4.7.2 Stress corrosion cracking (SCC)

Hopefully, it can be seen that Zr-based alloys will corrode slowly in water-based coolants; we discussed previously that due to fuel expansion and interaction with cladding we now have points of increased stress. As you can probably imagine, increased stress and corrosion may not go well together, and indeed you would be correct. The increased stress, aligned with corrosive media, in this case water, increases the corrosion, and this is known as SCC, an example is shown in figure 4.16. This process has multiple implications for the use of Zr-based alloys in reactors, but by far the largest is the possibility of the fuel itself, normally within the clad, being in contact with the coolant.

This can change the nature of the fuel dramatically, giving rise to new phases/oxides being formed:

$$UO_2 + xH_2O \rightarrow UO_{2+x} + xH_2. \tag{4.3}$$

The formation of UO_{2+x}, i.e. hyperstoichiometric UO_2, leads to the modification of properties detrimental to stability and, ultimately, can lead to the formation of UO_3. The formation of H_2 can dissolve in the liquid coolant, bubble away as gas, or, more insidiously, dissolve into the Zr-clad. Each of these products is disadvantageous to long-term use of the fuel/cladding, shown in figure 4.17.

4.7.3 The role of zirconium hydrides

As we discussed in chapter 3, ZrH/ZrH_2 is detrimental to the long-term stability of Zr-based alloys. Since the cladding is now placed into a medium that will increase the degree of hydridation, does this affect the long-term stability of the cladding? The simple answer to this question is yes, primarily for the same reasons as outlined before: increased brittleness, crack formation, etc, an example is shown in figure 4.18. When it links with PCI and SCC, the impacts can be magnified and lead to premature failure in the cladding.

So far we have only discussed the corrosion in terms of pure chemical corrosion, but once the effects of radiation from the core are included, it becomes more complex. For every chemical process we have in the corrosion, and for every

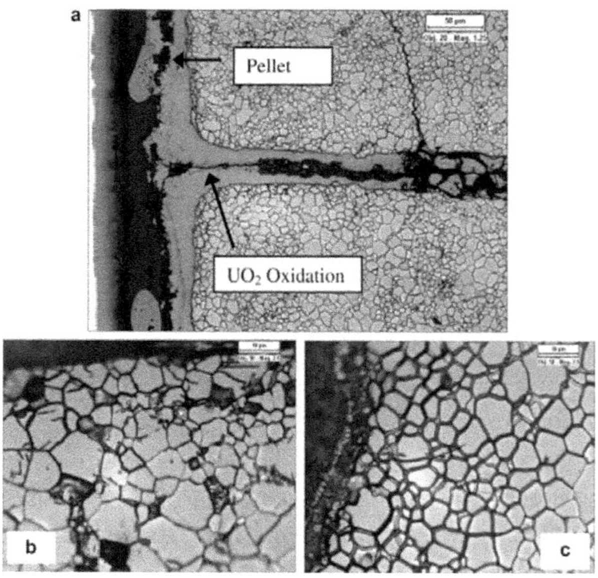

Figure 4.17. Image of pellet oxidation after PCI/SCC [10].

Figure 4.18. Cross-section of cladding, showing ZrH/ZrH_2, a UO_2 pellet and the corrosion layer on the surface [2].

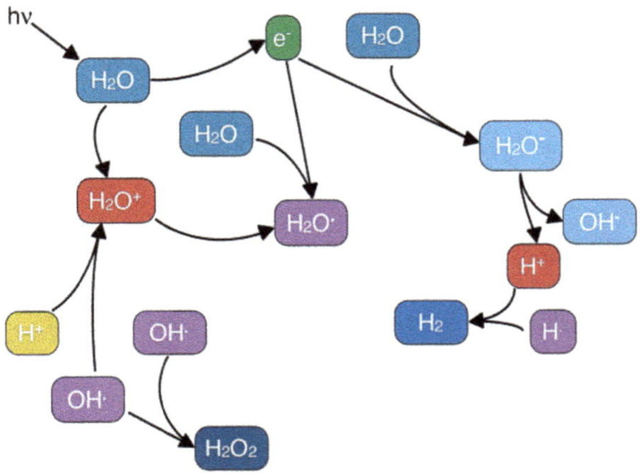

Figure 4.19. Schematic of radiolysis in water and the products.

radiation damage effect that can arise, there are further complications that can be added. In essence, the operational behaviour of fuel/cladding is highly complex.

4.7.4 Radiolysis of cooling water

Before we discuss the effects of radiolysis, we should define what it is and where it comes from. Radiolysis in the simplest terms is the breakdown of H_2O through interactions with radiation. Now consider the interior of a fission core; here, in addition to the splitting of the atoms, we have extremes in radiation, for example, gamma ray formation. This radiation interacts with the water, forming a range of species, from splitting the molecule into H^+/OH^- to forming radical forms of water, e.g. H_2O. These new species interact with whatever is present in the water, increasing the corrosion. In figure 4.19, some of the potential corrosion pathways for water that has been 'activated' from being within the reactor core are shown.

So, if using Zr-based clad and H_2O coolant is so problematic, why do we use them? The answer to this question is both complex and easy to answer; in an ideal world, we would not use them, but we have extensive experience in this area and can now predict/mitigate many of the effects we have discussed. However, this does not mean that alternative technologies have not been invented, and these are the topic of the next chapter, where we discuss alternative fuel/reactor designs.

References

[1] Rondinella V V and Wiss T 2015 The high burn-up structure in nuclear fuel *Mater. Today* **13** 24–32

[2] Olander D 2009 Nuclear fuels—present and future *J. Nucl. Mater.* **389** 1–22

[3] Chadwick M B *et al* 2006 ENDF/B-VII.0: next generation evaluated nuclear data library for nuclear science and technology *Nuclear Data Sheets* **107** 2931–3060
Chadwick M B *et al* 2011 ENDF/B-VII.1 nuclear data for science and technology: cross sections, covariances, fission product yields and decay data *Nuclear Data Sheets* **112** 2887–996

[4] Nicoll S, Matzke H J, Grimes R W and Catlow C R A 1997 The behaviour of single atoms of molybdenum in urania *J. Nucl. Mater.* **240** 185–95

[5] van Uffelen P, Konings R J M, Vitanza C and Tulenko J 2010 Analysis of reactor fuel rod behaviour *Handbook of Nuclear Engineering* ed D G Cacuci (New York: Springer)

[6] Hagrman D L, Reymann G A and Mason R E 1981 *MATPRO Version 11 (Revision 2): A Handbook of Materials Properties for Use in the Analysis of Light Water Reactor Fuel Rod Behavior* vol Rev 2 *NUREG/CR- 0497 TREE-1280* Rev 2

[7] Ding H, Virkar A V and Liu F 2012 Defect configuration and phase stability of cubic versus tetragonal yttria-stabilized zirconia *Solid State Ionics* **215** 16–23

[8] White J T and Nelson A T 2013 Thermal conductivity of UO_{2+x} and U_4O_{9-y} *J. Nucl. Mater.* **443** 342–50

[9] Byers W A 2010 Pressurized water reactor core crud mapping *18th International Conference on Nuclear Engineering (ICONE 18, Xian, China) ASME Proc.* **5** 333–9

[10] Higgs J D, Lewis B J, Thompson W T and He Z 2007 A conceptual model for the fuel oxidation of defective fuel *J. Nucl. Mater.* **366** 99–128

IOP Publishing

Nuclear Materials Science

Karl Whittle

Chapter 5

Evolution of reactor technologies

So far we have discussed the material effects of radiation damage within a reactor, the changes that occur within fuel during operation, and the choices made for fuel containment, i.e. cladding. However, how these have modified reactor designs is a topic we have not considered. This chapter takes an evolutionary look at reactor design, picking up on the key innovations and how they have evolved reactor design.

Before we discuss the early reactors, we need to clarify one of the key criteria for a reactor that was mentioned in the introduction, i.e. the transformation of energy from kinetic energy to electricity using a Carnot cycle, as shown in figure 5.1. There are currently two main methods for this, both of which use the conversion of heat energy to kinetic energy to drive turbines. One uses a gas, such as CO_2/He, and is called the Brayton cycle, and the other uses steam and is known as the Rankine cycle. For the purposes of this discussion, we will not cover the processes involved, as primarily they are not essentially nuclear materials related or limited, and in addition they follow standard thermodynamic concepts, which are better described elsewhere. In this chapter we will discuss the material requirements for the conditions used within the core, e.g. the ~873–923 K CO_2 used in AGR gas turbines.

As has been covered many times already, the key requirement for operational use within a reactor is that any material must allow the fission process to continue, and as a result certain elements/materials cannot be used within the core. We have also discussed how material choices can be advantageous to the fission process, and how they can be mitigated if they are deleterious. For example, slowing neutrons is a recognised method for increasing fission yield, and using CO_2 or H_2O allows for this. At the same time, we can use both CO_2 and H_2O as a method for removing heat from the core. Thus in both LWR and AGR reactor designs we have a material that can assist reactor operation in two areas. It is how these advantages have been reached that concerns us in this chapter, and this also provides an introduction to the next chapter, which concerns the challenges associated with the next generation of reactors.

doi:10.1088/978-0-7503-1104-5ch5

Figure 5.1. Example of the Carnot cycle converting kinetic energy (wind) to electricity, see also animation.

Figure 5.2. Sketch of the reactor at Stagg Field, Chicago. Reproduced with permission from the Argonne National Laboratory.

5.1 Generation I—prototype reactors

It is a recognised truth that early reactors were not always designed with civil applications in mind, but were designed with a military focus. However, what was learnt during this time proved to be valuable and assisted civil reactor design greatly. Two early reactors, both part of the Manhattan project, paved the way for civil nuclear power, it was just not the primary focus at the time. For example, the reactor that was famously situated in a squash court at Stagg Field at the University of Chicago (figure 5.2) was the first manmade nuclear fission reactor, using a graphite core to moderate the neutrons and increase fission. What was key at the time was that the graphite contained significant amounts of boron, which, as we have discussed, absorbs neutrons, reducing the likelihood of sustainable nuclear fission.

Figure 5.3. Photograph of the X-10 reactor. Reproduced with permission from American Museum of Science and Energy.

This led to the development of boron-free, or reactor-grade graphite, which is key to the operation of CO_2-cooled reactors, such as the AGR.

The second early reactor of interest is X-10 (figure 5.3), based at what was then known as the Clinton Engineer Works and is now Oak Ridge National Laboratory, which provided the early plutonium to Los Alamos. This reactor evolved the Stagg Field design into a larger one and was air-cooled. In this design the fuel is loaded into slugs, which are placed into the core, and once irradiation has been completed, fresh slugs are pushed in, pushing the irradiated fuel out, as described in figure 5.4. This method was also used at the Windscale reactor in the UK, which we will discuss in a later chapter. The key evolutionary step was not materials focused *per se*, but came from the extraction of the formed ^{239}Pu, using a chemical process based on bismuth phosphate, which has subsequently evolved into the methods used today for reprocessing spent nuclear fuel.

The third key reactor during this time was based at Hanford, Hanford B (figure 5.5), which produced the plutonium used in the Trinity test and at Nagasaki. The design of Hanford B focused on the transmutation from ^{238}U to ^{239}Pu, or breeding. In order to achieve maximum efficiency, an evolved design was used, giving rise to two evolutionary steps. The first was to use water to cool the reactor, in this case from the nearby Columbia river, with this passing through the reactor and into a settling pond before being allowed back into the river (figure 5.6) [1]. As water was now being used, the fuel had to be contained; think back to the

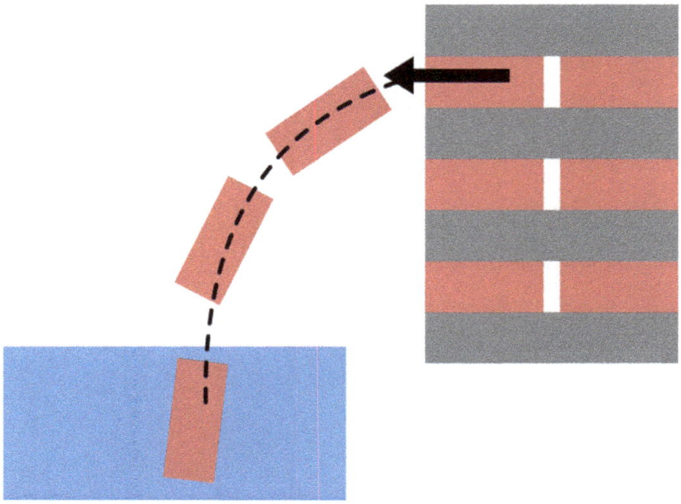

Figure 5.4. Process by which the fuel was introduced and ejected in X-10, see also animation.

Figure 5.5. Photograph of the Hanford B reactor during construction. Reproduced with permission from [1].

effects of water on U and UO_2/UO_3 in previous chapters, and the methods described. In Hanford B, the fuel was uranium metal contained with aluminium cans, an important step and the earliest form of operational cladding. As such, Hanford B is sometimes considered a prototype for modern reactor designs.

Thus, at the end of Manhattan project, we can see the design ideas that are still followed today:

(i) design of the core to allow continued fission chain reaction;
(ii) cooling of the core to allow extended use and prevent melting of the fuel;
(iii) containment of the fuel to allow fission to continue, whilst preventing corrosion damage to the fuel.

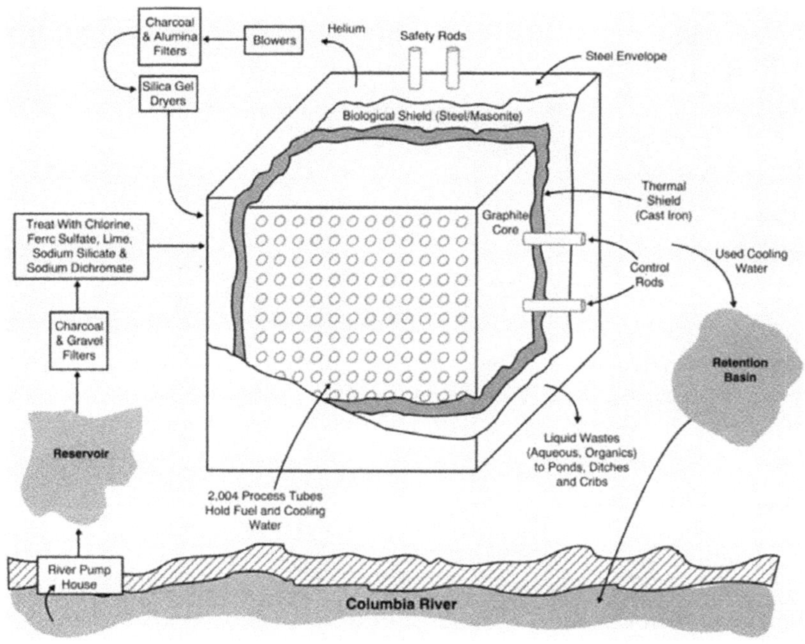

Figure 5.6. Schematic of the cooling of the Hanford B reactor. Reproduced with permission from [2].

As an aside, the terms developed by the engineers during this early work are still used today, and two are used regularly.

i) The unit of 'barn' (10^{-28} m^2) is used to represent neutron cross-sections, whether they be absorption or fission. The barn is one in a series of units, and it is said to have been based on the phrase 'as big as a barn'. Interestingly, two further units are the 'outhouse' and the 'shed', which are 10^{-34} m^2 and 10^{-48} m^2, respectively.
ii) SCRAM refers to an emergency reactor shutdown, and it has two possible origins:
    i) it could be from the phrase 'Let's scram out of here', or
    ii) 'Safety cut rope axe man/single control rod axe man'—coming from the early control rods at Stagg Field being held by a rope above the reactor, which, if necessary, would be cut with an axe.

Both are still used today, and SCRAM was used to identify the emergency shutdown controls in early reactors (figure 5.7).

After the Manhattan project, it was accepted that nuclear reactors could be used for power generation. This lead to programmes to develop reactors not only for large scale civil power, but also for naval/air applications. These were also the first reactors to utilise the heat from the core through a Rankine or Brayton cycle to generate electricity.

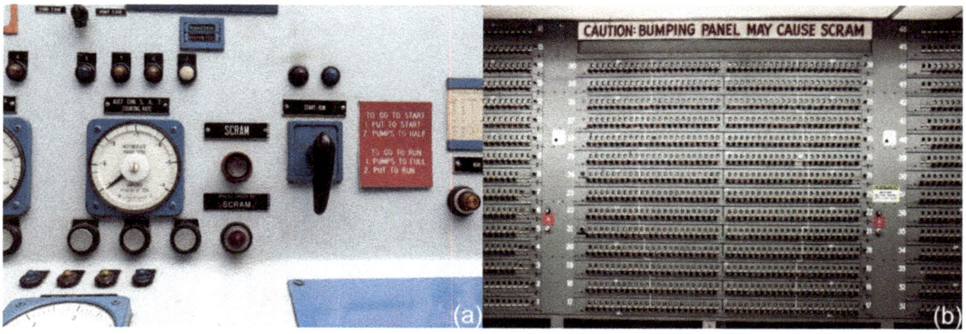

Figure 5.7. (a) The emergency shutdown button in a reactor. This NS Savannah SCRAM MD1 image has been obtained by the author(s) from the Wikimedia website where it was made available by Acroterion under a CC BY-SA 3.0 licence. It is included within this article on that basis. It is attributed to Acroterion. (b) A wall monitoring pump at Hanford. © 2008 Stuart Isett/www.isett.com (a) and (b) both show the word SCRAM.

5.1.1 Pressurised water reactors (PWR)

In 1947, Admiral Rickover was appointed to oversee the US naval reactor programme, and it is at this time that the modern themes of reactor design can start to be drawn. For example, the acceptance that operational efficiency is key to a reactor's sustainability. In the 1950s there were two competing naval reactor designs for use within a submarine, a PWR powering the USS Nautilus, launched in 1954, and a sodium-cooled reactor powering the USS Seawolf, launched in 1957 [3]. These reactors each had their advantages, for example the USS Seawolf reactor required 40% less space than the PWR in the USS Nautilus because of the different (electromagnetic) method of pumping used. However, as with most things, while there were advantages, there were also disadvantages, and in the case of the USS Seawolf, there were significant disadvantages, not the least of which was the decay of ^{24}Na from neutron activation within the coolant. During the developmental phase there were so many problems that the USS Seawolf was converted to a PWR in 1958. At the time, Admiral Rickover was quoted [4] as saying that the reactor was, 'expensive to build, complex to operate, susceptible to prolonged shutdown as a result of even minor malfunctions, and difficult and time consuming to repair'.

During this programme, a commercial nuclear power plant was proposed by President Eisenhower, and given to Admiral Rickover to oversee. At the time, the only suitable reactor design was the PWR, developed from the version used in the USS Nautilus and proposed for an aircraft carrier. The facility, known as the Shippingport atomic power station (figure 5.8), was built in Shippingport near Pittsburgh, PA, and became operational in December 1957, running for 25 years. During the operation of the reactor, there were three major cores, each providing valuable information for the development of reactor technology. For example, in the first core the basics of refuelling were utilised to ensure efficient operation, and enriched fuel was used, whereas the final core developed the basis for Th-based fuel. In this final core, which used predominantly ^{232}Th with a ^{233}U seed, it was discovered after five years of operation that ~1.5% more fissile material was present than at the beginning, indicating the potential for using ^{232}Th as a fuel [4].

Figure 5.8. The Shippingport pressure vessel during installation. Reproduced with permission from [5].

During this time, the move towards using Zr-based alloys for clad began, and it continues today— see the previous chapters for why Zr is used. For example, the use of Zircaloy-4 in PWR and Zircaloy-2 in BWR designs began, and related alloys such as Zirlo and M5 are used today.

Thus from the experiences of the early naval submarine and Shippingport reactors we can see the genesis of modern reactor designs based on pressurised water technology, with the use of steel for the reactor pressure vessel. However, a reactor design that used a different basis was developed in the UK (the first to generate commercial power), and this had implications for future reactor designs.

5.1.2 Gas-cooled reactor (MAGNOX)

The design basis for a MAGNOX reactor is different to that for a PWR; rather than using water as the coolant, CO_2 gas is used, which then drives the turbines, i.e. a Brayton cycle. The MAGNOX took a different design approach for cladding: as outlined in chapter 3, Mg has a lower neutron absorption than Zr, but also a lower melting point. In the case of MAGNOX, the reactor was designed to operate at ~360° C, and since Mg reacts with water, CO_2 was used as the cooling medium. A secondary difference in design compared to a PWR is the fuel cladding, in a MAGNOX there are fins which increase surface area, and therefore conduction into the gas. A schematic is given in figure 5.9.

The first MAGNOX reactor to generate energy was Calder Hall (figure 5.10), located at Sellafield in Cumbria. It opened on 27 August 1956 and was the world's first power station to generate commercial energy from nuclear power [6]. The designed output of the MAGNOX reactors was 60 MWe from each core, coupled to standard steam turbines. During the operation of the reactor it became apparent that some of the reactor components, steel based, slowly corroded in the CO_2 gas,

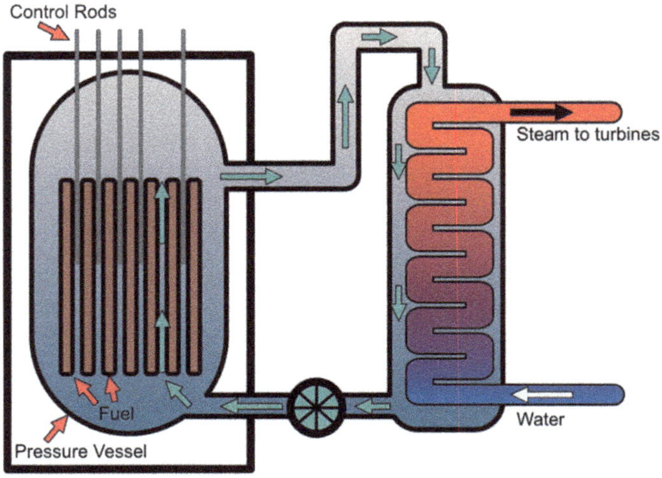

Figure 5.9. Schematic of a MAGNOX reactor (left) and fuel elements with fins that aid thermal conductivity (right, reproduced with permission from the Nuclear Decommissioning Authority).

Figure 5.10. The Calder Hall power station being built. Reproduced with permission from the Sellafield sites.

requiring the reactor to run at lower temperatures than designed. In total, 26 reactors were built in the UK, running from 1956 (Calder Hall) until 2015, when the last, Wylfa, was shut down.

During this early period in reactor development, there were other notable reactor designs, such as the boiling water design proposed by Idaho National Laboratory and General Electric. This design has some advantages when compared to a PWR, and has been developed continuously to form the basis for the ABWR, which we will discuss later in this chapter.

These reactor designs are commonly referred to as GenI reactors, and they led to the next generation of reactors, many of which are still in use today.

Table 5.1. Summary of common GenII reactor designs and example operators/reactors.

Reactor type	Countries	Moderator	Coolant	Example reactors
PWR	USA, France, Russia	H_2O	Liquid water ~575–585 K	Three Mile Island, Sizewell B
BWR	Japan, Sweden	H_2O	Liquid water/steam ~540 K	Fukushima
PHWR	Canada, India	D_2O	Liquid D_2O ~575–585 K	NRU
AGR	UK	C	CO_2 ~920 K	Wylfa, Trawsfynydd
RBMK	Russia	C	Liquid water/steam ~540 K	Chernobyl

5.2 GenII—commercial reactors

Whereas the MAGNOX and Shippingport reactors were designed to supply power, they also had the ability to breed fissile material. For example, two of the MAGNOX reactors were modified to permit plutonium extraction if needed, and indeed the business case for the MAGNOX reactors had a plutonium buy-back provision from the UK government. In the case of the next generation of reactors, commercial operation was the designed goal. The common designs, outside of the Soviet Union, were principally evolutions of existing technology. The major reactor types are shown in table 5.1.

The designs outlined in table 5.1 are generally evolutionary designs of GenI designs, such as the PWR, BWR or MAGNOX. The exception to this was the RBMK, which was designed to be as cheap and efficient as possible, capable of using natural uranium, and use graphite moderation with light water cooling. However, whilst this design did achieve it desired goals, it did have noted design issues, and was the reactor design used at Chernobyl, which we will discuss in more depth in a later chapter.

The key difference between GenI and GenII designs is increased awareness of safety and risk. Reactor designs now included methods by which previously potential identified problems could be dealt with, or designed out. An example is the addition of soluble neutron absorbers to the cooling system in the event of an emergency, to prevent runaway fission. One area where the PWR/BWR have an advantage over the RBMK/CANDU designs is the prevalence of a negative void co-efficient, which helps to keep the reactor from becoming unstable. So, what is a void-coefficient?

In a cooling system base which has the ability to be either liquid or vapour—e.g. H_2O, which can be water or steam—there is the possibility of two different neutron absorptions, based on density, i.e. bubbles within liquid, and this is the void coefficient. This void coefficient can have one of two effects, as voids form within the reactor can either increase or decrease output. When a reactor is said to have a positive void coefficient, it increases output with increasing void formation, whereas for a negative coefficient the opposite is true. Generally, PWR and BWR have a negative void coefficient, whereas the RBMK and CANDU (a PHWR design) have a positive one. Thus PWR and BWR would be expected to have decreased activity with increasing void formation. The RBMK initially had a relatively large coefficient, which contributed to the problems at Chernobyl [7]. Gas-cooled reactors do not have this problem, and are said to be inherently safer.

So, what are the current key material challenges in GenII reactors? As with the designs initiated at Shippingport and Calder Hall, the issues of temperature and corrosion remain. For example, while Zr-based alloys remain prevalent in LWR technologies, the alloys themselves have been modified to extend the lifetime in the reactor and improve the efficiency of the reactor.

One of the key developments from GenI to GenII was the AGR, developed in the UK. These are important, as they link to future GenIV reactor designs.

5.2.1 Advanced gas-cooled reactors (AGR)

These reactors introduced new challenges for reactor materials, with an increase in designed thermodynamic efficiency, the coolant output temperature was raised from ~625 to ~925 K, with CO_2 again being used as coolant [8]. However, such a rise in temperature moves the system away from using Mg-based alloys as cladding, as they would not withstand the temperatures. Initially, Be was proposed for use as cladding, however, this proved to be unsuitable for use and as a consequence a stainless steel alloy was chosen (see chapter 3). Since the AGR now used cladding that would absorb a significant number of neutrons, enriched fuel was used, similar to that used in PWR and BWR designs, a schematic is shown in figure 5.11.

Since the AGR operated at elevated temperatures compared to their PWR/BWR brethren, metallurgical processes that would be slow in light water reactors could now be problematic. If you remember, when discussing the effects of radiation

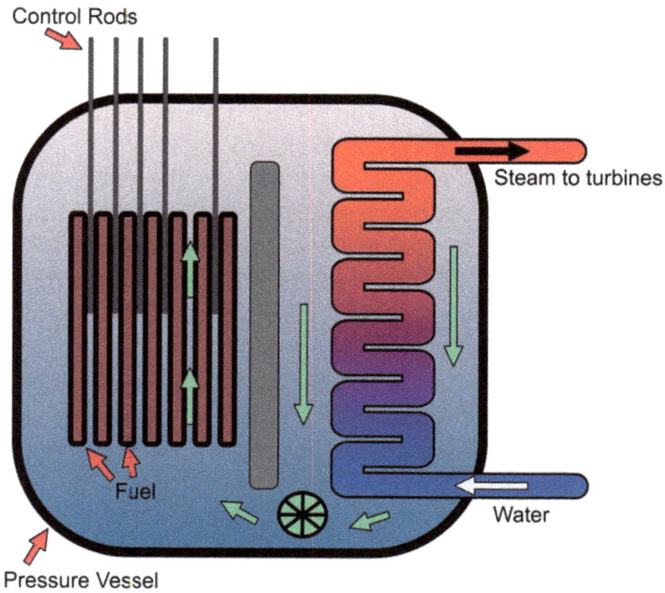

Figure 5.11. Schematic of the AGR design.

Figure 5.12. Schematics for (a) Westinghouse AP [9], (b) Areva EPR [10] and (c) Hitachi ABWR reactor designs [11].

damage, we went through the main problems, such as cracking and creep. While they occur in light water reactors, from a thermodynamic viewpoint, the likelihood of them occurring in an AGR increases. Hence, during the operational lifetime of the AGR reactor fleet, a great deal of information on materials experiencing high-temperatures and radiation effects has been collated, for example regarding the effects of damage on graphite. For example, one design improvement made to the AGR compared to the MAGNOX is that the boilers are now placed within the core, and not outside. Whilst it may appear that the AGR design was an evolutionary dead end, it does provide information that is useful for future designs, such as GenIV.

As the GenII designs were predominantly an evolution of GenI designs, the same can be said of the GenIII/GenIII+ designs that form the current generation of new reactors being built.

5.3 GenerationIII/generationIII+—evolved designs

Generally speaking, GenIII/GenIII+ reactors are evolutions of previous designs, with enhancements in the following areas:

(i) enhanced power output;
(ii) evolved safety systems and accident prevention;
(iii) increased utilisation of fuel;
(iv) extended lifetime for design.

Each of these areas aimed to increase the economic and societal acceptance of nuclear energy. Rather than going through each of the GenIII/GenIII+ designs (GenIII+ are evolved GenIII designs) shown in figure 5.12, we will focus on the three most common designs proposed for use, and how they compare with previous designs and each other.

(i) Westinghouse AP;
(ii) Areva EPR;
(iii) GE-Hitachi ABWR.

5.3.1 Westinghouse AP and Areva EPR

As shown in both the introduction and the discussion of GenI and GenII designs, this reactor design uses pressurised water, operating at ~585 K, flowing around the core. This cooling circuit transfers heat to the secondary cooling circuit, which in turn heats water to steam, driving turbines, i.e. a Rankine cycle. The key changes from previous models are in the power output, expected operating lifetime and enhanced safety systems.

Previous PWR designs had a range of outputs, for example TMI-1's (Three Mile Island) was ~850 MWe, while Sizewell B's is ~1250 MWe. The Westinghouse AP design provides for an output of 1300 MWe [9], but with a smaller footprint when compared with previous designs, whereas the EPR design provides for outputs up to 1600 MWe [10]. The reactors still use a zirconium-based fuel cladding, Zirlo for the AP and M5 for the EPR, both proprietary alloys developed by the reactor manufacturers, and both using low enriched UO_2 for fuel. Both reactor designs have multiple redundancy in safety systems, which range from the passive to active, for example in both designs the control rods are gravity assisted, i.e. they fall in from the top, so that in the event of a power failure they can still be inserted. There are, however, two major differences in design that are worth considering.

The Westinghouse AP design has one, currently unique, design addition in the event of a loss of coolant accident (LOCA), which, as the name suggests, is when the core has no cooling water. At the top of the reactor building there are tanks filled with water, which are designed to rupture in the event of a core breach with a release of heat. Once ruptured, the water will fall down into the core area and evaporate, taking the heat with it. As the steam rises, convection currents will come into contact with the exterior surface of the building, condensing the steam back into water, which will flow/fall to the bottom of the building, where it will turn into steam (figure 5.13). Such a cycle should help to keep the reactor cool while the active cooling systems are brought back online [9].

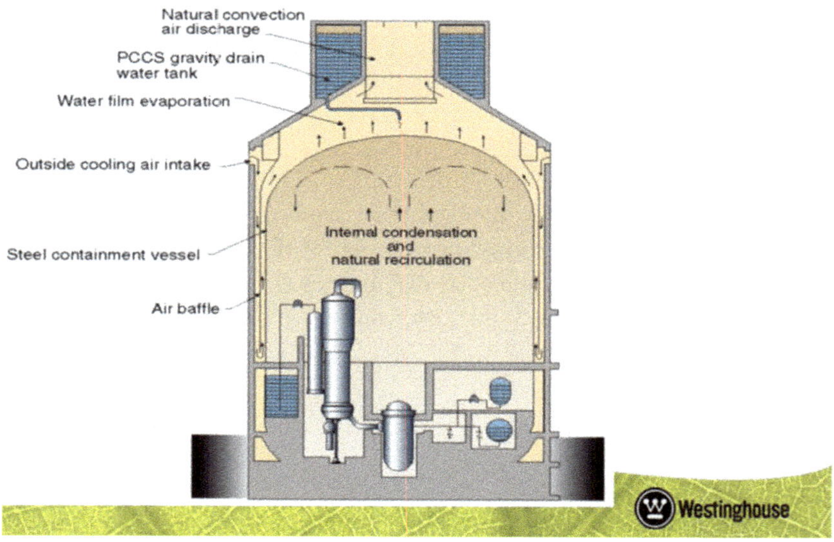

Figure 5.13. Representation of the proposed cooling process during an accident within a Westinghouse AP reactor. Reproduced with permission from [9].

The Areva EPR design takes a different, but equally valid, approach—here there are multiple cooling systems available, with an increased level of redundancy. However, if needed, there is a pit that can be used to empty a molten core, increasing the surface area available for cooling [12]. Both this and the Westinghouse approach are designed to be used in the last resort, and not to be a normal part of reactor operation.

5.3.2 GE-Hitachi ABWR

This design, like the PWR designs, is an evolution of previous BWR reactors, now incorporating enhanced safety systems. As with PWR evolution, there is an increase in power output to ~1350 MWe, still with zirconium-based fuel cladding and low enriched UO_2 as the initial fuel. One key advantage is the designed ability for a pad beneath the core to capture and contain a melted core. This pad has a separate cooling system that can be used to cool the molten core and contain the melt. This is similar to what is proposed in the EPR PWR design and highlights that good designs are often incorporated elsewhere [13].

5.3.3 General design considerations

One of the key improvements in GenIII/GenIII+ designs is the increased utilisation of fuel. If you remember back to where we discussed the operational effects of fission on both fuel and cladding, such an increase would give rise to issues with long-term stability in the materials. Hence, increasing the burn-up within fuel where possible is highly desirable as a means to increase fuel efficiency. There is, however, the possibility of using MOX-based fuel, which, as was discussed earlier, is a mixture of both UO_2 and PuO_2. All of the current GenIII/GenIII+ family of reactors can use such fuel, thus meeting the objective of enhanced use of fuel.

In previous generations of reactors, the expected operational lifetime of the reactor was assumed to be up to 40 years, and in some extreme cases they were designed for 50 years of use. For example, the designed lifetime of TMI-1, a Babcock & Wilcox design, was initially 40 years [14]. The current design criterion for the majority of GenIII/GenIII+ designs is 75 years, before any extensions are required. This extended time of operation can cause issues, not for the regularly changed parts of the core, such as fuel assemblies, but for those that cannot be changed, such as the pressure vessel. The bulk effects of radiation damage, such as cracking and creep, now need to be monitored and, if necessary, mitigated/ameliorated.

As has been outlined, the evolution of reactor design has led to changes in material use and the exploitation of materials development to enhance reactor operation. For example, modifications to Zr alloy composition (chapters 3 and 4) have greatly enhanced reactor operation. The next step in reactor evolution, the GenIV designs, challenges both reactor designers and those developing the materials required for their manufacture.

There is, however, one reactor design that we have not yet discussed, which has major implications for future GenIV design. This design had its genesis during the

GenI and GenII designs and, if decisions had gone another way, could be the basis for the reactors being used today. This reactor design uses a molten salt rather than water or gas as the coolant/moderator, initially in a unique situation.

5.4 Molten salt reactors

An alternative approach to running nuclear reactors was proposed and tested in the 1950–60s, based on the use of molten salts for fuelling and cooling the reactor. Such a design has many potential advantages, for example, running water-based reactors limit the temperature of operation (< 525 K), whereas molten salt reactors can run much hotter, increasing thermodynamic efficiency. A second equally advantageous property of using molten salts as opposed to H_2O, is the lower vapour pressure at temperature, reducing the risk of a pressure-induced leak. Finally, using a molten salt-based cooling system allows a liquid fuel embedded within to be used. This final advantage provides an opportunity for both online refuelling and continuous purification of the salt, removing fission products and preventing some of the issues associated with fission product formation.

However, there is one disadvantage, increased corrosion from the use of molten salts in the containment being used.

5.4.1 The molten salt reactor experiment (MSRE)

One of the earliest molten salt reactor designs was at Oak Ridge National Laboratory, running from 1965–69, the Molten Salt Reactor Experiment (MSRE). The design used a liquid fuel/coolant mixture of $LiF:BeF_2:ZrF_4:UF_4$ in the ratio 65:29:5:1 as the primary circuit. In a manner similar to a PWR, it had a secondary cooling circuit of $LiF:BeF_2$ in the ratio 2:1 [15]. To moderate the fission reaction, a graphite core was used, and a nickel-based alloy, Hastalloy-N, was used for reactor containment. As this was a test reactor, and no attempt to generate electricity was made, the core was air cooled, operating at ~950 K, with gas blowing over radiators, shown in figure 5.14. The programme finished in 1969, with no further development proceeding, because of multiple factors which could have been overcome with time. However, by this time light water reactors were becoming prevalent and the material challenges, specifically corrosion and radiation damage, were problematic [15]. Whilst the MSRE was one of the early experimental molten salt reactors, it was not the first, that privilege goes to a unique design proposed for a unique application—powering a jet plane in flight.

5.4.2 Aircraft reactor project

Before the MSRE programme, a proposal was put forward by the US Air Force in 1946 for a plane that could fly indefinitely using nuclear-based propulsion. The reactor was designed around a primary circuit of $NaF:ZrF_4:UF_4$ in the ratio 53:41:6, using BeO as a moderator and liquid Na in a secondary cooling circuit, and shown in figure 5.15. The reactor was designed to operate at a temperature of 1130 K and ran for 1000 hours. This reactor would then be used to power the jet plane in one of two ways, directly and indirectly [16, 17].

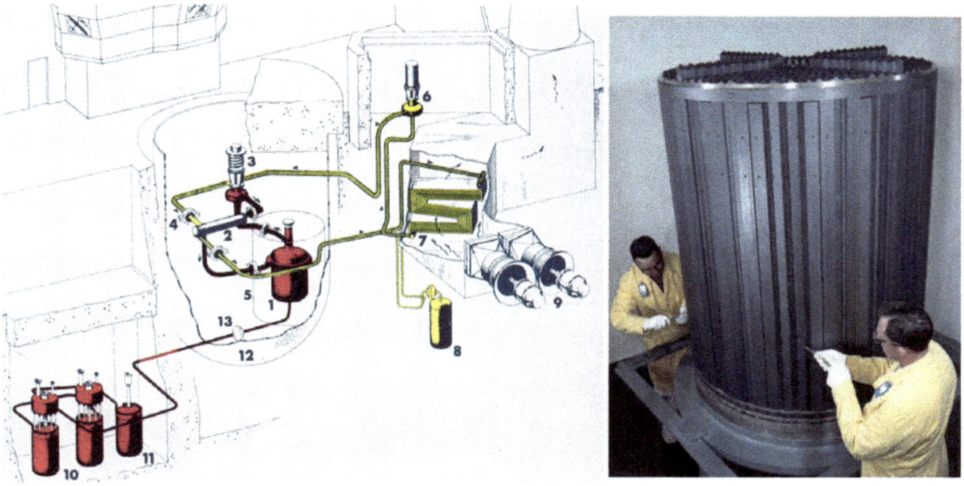

Figure 5.14. Schematic of the molten salt reactor (a) and photograph of the cooling system (b).

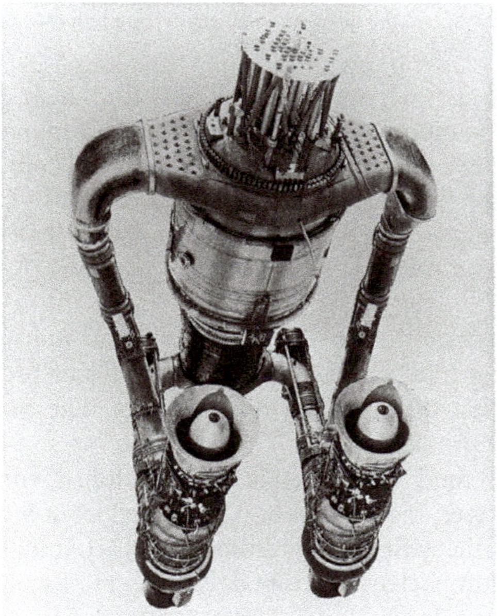

Figure 5.15. Proposed design for the reactor in the aircraft reactor project [16, 17].

Before we describe the proposed methods, we should summarise how a jet engine operates. In very simplistic terms, a jet engine compresses air, which is heated and ejected, driving the engine forward. Currently, jet engines use fuel to heat the air, but what if we use another source of heat, could we achieve the same effect? This was the design concept behind the aircraft reactor project, and two methods were proposed during the testing phase, direct and indirect heating of the air.

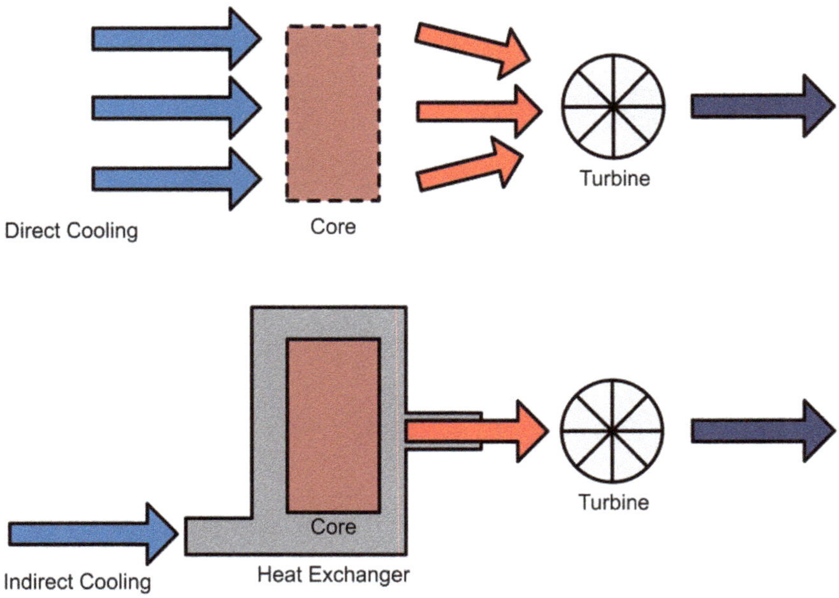

Figure 5.16. Schematic of the direct and indirect heating proposed for jet engine use.

- Direct heating—as the name suggests, this took heat directly from the core and heated the compressed air. This worked in a manner similar to a BWR, with direct heating of the air passing through the core (figure 5.16).
- Indirect heating—this operated in a manner similar to a PWR, passing the air through a heat exchanger, a liquid Na loop, which then heated the air prior to it entering the compressor.

The direct design was tested in the laboratory and did indeed fly within a jet aircraft (47 times for a total of 89 hours), but it did not provide power, and was rather was flown to test the shielding requirements. The indirect design was less successful and proved problematic.

The programme ran until 1961, when it was abandoned without ever powering an aircraft in flight. However, the information generated from it was subsequently used in the development of the MSRE programme at ORNL and, as with the submarine reactor programme, it impacted on future development of reactors and the materials used within.

5.5 Summary

In this chapter we have examined the history of reactor design and can draw some conclusions on how the choice of design is impacted on by the choice of material.

5.5.1 Early reactors (GenI)

These reactors were developmental designs with materials chosen that met the desired criteria being addressed at the time. For example, the Hanford reactor B1 was designed to breed Pu for the Manhattan project and introduced the concept of cladding, while the naval liquid sodium reactor was discontinued, as it proved to be unreliable, leaving the PWR as the focused design.

5.5.2 Commercial reactors (GenII)

These expanded the concepts used in GenI designs and led to commercial power plants. For example, the AGR design implemented the lessons learnt from the materials and design of the MAGNOX reactors, increasing both the temperature of operation and power output. Light water reactors continued to develop during this time, leading to new Zr-based alloys with enhanced desirable properties. The majority of reactors currently operating worldwide are generally classed as GenII designs, with some running 40–50 years.

5.5.3 Evolved designs (GenIII/GenIII+)

The third generation of reactors are those being built now, and they have implemented the lessons learnt from GenI/GenII designs, showing enhanced output, operational life and safety. For example, the designed lifetime for reactors is now ~75 years, which could only have been achieved with continuous materials development during the previous generations.

However, now that GenIII/GenIII+ designs are being built, does this mean we have reached the ideal nuclear reactor technology? The probable answer to this is yes and no; yes, we have reached a safe technology, which is reliable, but no, it is not the most efficient. Are there are other reactor technologies being proposed which improve on GenIII/GenII+? The answer to this is yes. Whilst it can be argued that GenIII/GenIII+ designs are now mature, with further developments being incremental, e.g. increases in power, we are still limited in the operating temperature. This temperature is limited by the method of cooling used—water, which cannot be used above 650 K, where it is always a gas. As we have seen when using molten salts or gas for cooling, the potential operating temperature is much higher.

The driving force for the next generation of reactors, GenIV, is to move away from liquid water-based cooling and develop reactors that have increased output, efficiency and fuel utilization, and offer increased proliferation resistance. In the next chapter, we shall look at the GenIV reactor technologies, and some of the other alternative designs being proposed, and describe the material challenges associated with them—in some designs, the challenges are not trivial to overcome.

References

[1] FG Gosling 2010 The Manhattan Project: Making the Atomic Bomb (DOE/MA-0002)

[2] US Department of Energy The Manhattan Project. An Interactive History www.cfo.doe.gov/me70/manhattan/final_reactor_x-10.htm

[3] Hewlett R G and Duncan F *Nuclear Navy: 1946–1962* (Chicago, IL: University of Chicago Press) p 274

[4] Olsen G L, McCardell R K and Illum D B Fuel Summary Report: Shippingport Light Water Breeder Reactor (INEEL/EXT-98-00799 Rev 2) (Idaho Falls, ID: Idaho National Laboratory)

[5] US Department of the Interior (National Park Service) Survey number HAER PA-81-87 *Historic American Engineering Record* (US Department of the Interior, National Park Service)

[6] Jaffrey R 2002 *Power Eng. J.* **2002** 186–92

[7] IAEA 2005 Accident analysis for nuclear power plants with graphite moderated boiling water RBMK reactors *Safety Reports Series* no. 24 (Vienna: IAEA)

[8] Ecob R C, Lobb R C and Kohler V L *J. Mater. Sci.* **22** 2867–80

[9] Westinghouse Nuclear *AP1000 Overview*

[10] Areva EPR *Reactor Fact Sheet*

[11] Hitachi-GE Nuclear Energy Ltd *Advanced Boiling Water Reactor (ABWR)*

[12] Nuclear Directorate *Generic Design Assessment—Step 2 Fault Analysis Assessment of EDF Areva Submission for the EPR*

[13] Office of Nuclear Regulation *Summary report of the Step 3 generic design assessment (GDA) of Hitachi-GE Nuclear Energy's UK advanced boiling water reactor (UK ABWR)* (Office of Nuclear Regulation)

[14] NRC *Three Mile Island 1—Pressurised Water Reactor*

[15] Briggs R B *Molten-Salt Reactor Program Semi-Annual Progress Report, ORNL3872* (Oak Ridge National Laboratory)

[16] Fraas A P and Savalainen A W *Design Report on the Aircraft Reactor Project, ORNL-2095* (Oak Ridge National Laboratory)

[17] Cottrell W B, Hungerford H B, Leslie J K and Meen J L *Operation of the Aircraft Reactor Experiment, ORNL-1845* (Oak Ridge National Laboratory)

IOP Publishing

Nuclear Materials Science

Karl Whittle

Chapter 6

The challenges for materials in new reactor designs

By far the greatest challenges for materials within a nuclear reactor are those found within the core, and generally the material choices made so far have done the job well. Given that reactors worldwide are predominantly based on water based cooling, we are limited in the temperature of operation (< 640 K). This prevents increased thermodynamic efficiency in the conversion of energy being released. To get round this problem, the next generation of reactors, GenIV, have been proposed, and these are predominantly based on non-water cooling, with some other reactor designs being put forward. Taking the concepts put forward in chapter 5, we will describe the material challenges going forward, and how they may be met.

6.1 Generation IV—genesis

In 2002, a roadmap was published that proposed future design concepts for nuclear reactors that achieved the following goals:

(i) sustainable energy generation with increased fuel utilisation;
(ii) minimisation of nuclear waste;
(iii) life cycle cost enhancement with a level of risk comparable to current technologies;
(iv) enhanced safety and reliability with low risk of core damage;
(v) increased proliferation protection.

The members of the grouping (the GenIV forum [1]) at that time were Argentina, Brazil, Canada, France, Japan, South Korea, South Africa, Switzerland, the United Kingdom and the United States. Since the initial report, further members have joined: China, Russia, Euroatom and organisations such as the IAEA and NEA. The initial proposed a set of metrics and design goals that the next generation of reactors should achieve, and proposed six designs, three using thermal neutrons and three fast.

doi:10.1088/978-0-7503-1104-5ch6

Table 6.1. Summary of key parameters for the GenIV reference designs.

Parameter	VHTR	MSR	SCWR	GFR	SFR	LFR
Power	600 MWth	1000 MWe	1700 MWe	600 MWth	1000–5000 MWth	3600 MWth
Fuel type	Solid	Solid/liquid	Solid	Solid	Solid	Solid
Coolant inlet/outlet	900/1300 K	850/1000 K	550/800 K	560/1120 K	620/820 K	~750 K
Coolant	He	F-based salt	H_2O	He	Liquid Na	Liquid Pb (Pb/Bi)
Carnot cycle	Brayton	Braying	Rankine	Brayton	Rankine	Brayton

These designs were:

(i) very high temperature reactor (VHTR);
(ii) molten salt reactor (MSR);
(iii) supercritical water reactor (SCWR);
(iv) gas-cooled fast reactor (GFR);
(v) sodium-cooled fast reactor (SFR);
(vi) lead-cooled fast reactor (LFR).

Before we can discuss what material challenges are expected across these designs, we should compare and contrast them.

6.2 Reactor types

6.2.1 Very high temperature reactor (VHTR)

This reactor is similar in nature to the AGR design we have already spoken about, with some key differences, i.e. it is gas cooled and grew out of previous gas-cooled high temperature reactors, such as DRAGON [2]. The initial focus of the reactor was not on generating electricity, but on using the heat produced for other applications, such as the generation of hydrogen. This has changed, such that electricity generation is part of the design, with the waste heat having other applications. The key criteria for the reference design are shown in table 6.1 and the schematic design is shown in figure 6.1.

6.2.2 Molten salt reactor (MSR)

The premise behind this design is similar, if not identical, to the original molten salt reactors in the 1950–60s. Indeed, the schematic design shows some comparable features. The key criteria for the design are shown in table 6.1 and the schematic design is shown in figure 6.2.

6.2.3 Supercritical water reactor (SCWR)

This design uses H_2O for heat transfer, but runs the reactor above 650 K, i.e. above the critical point for H_2O. In a fluid system the critical point is defined as being the

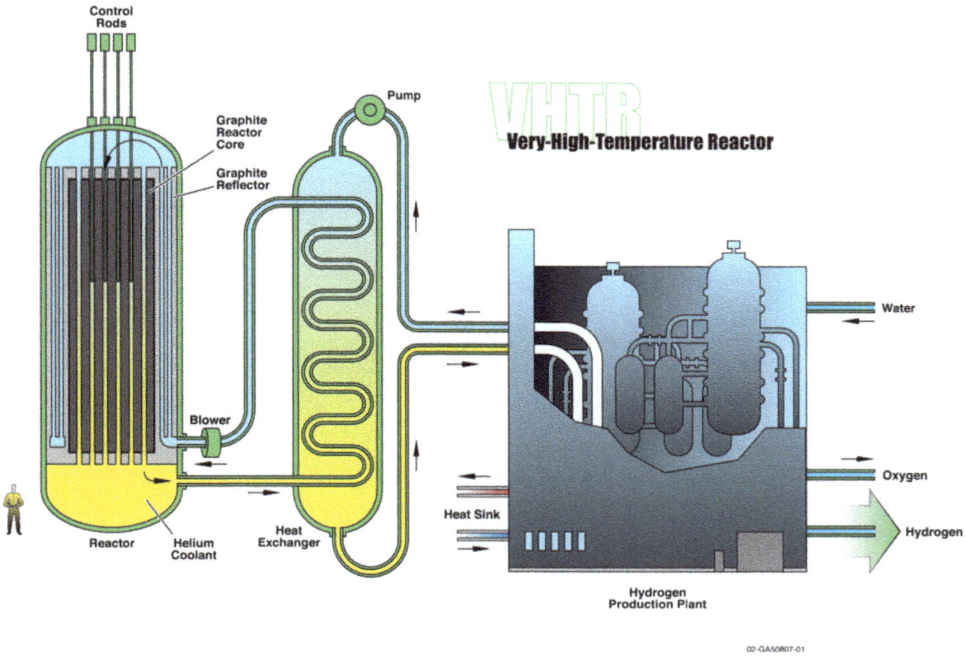

Figure 6.1. Reference design for the VHTR.

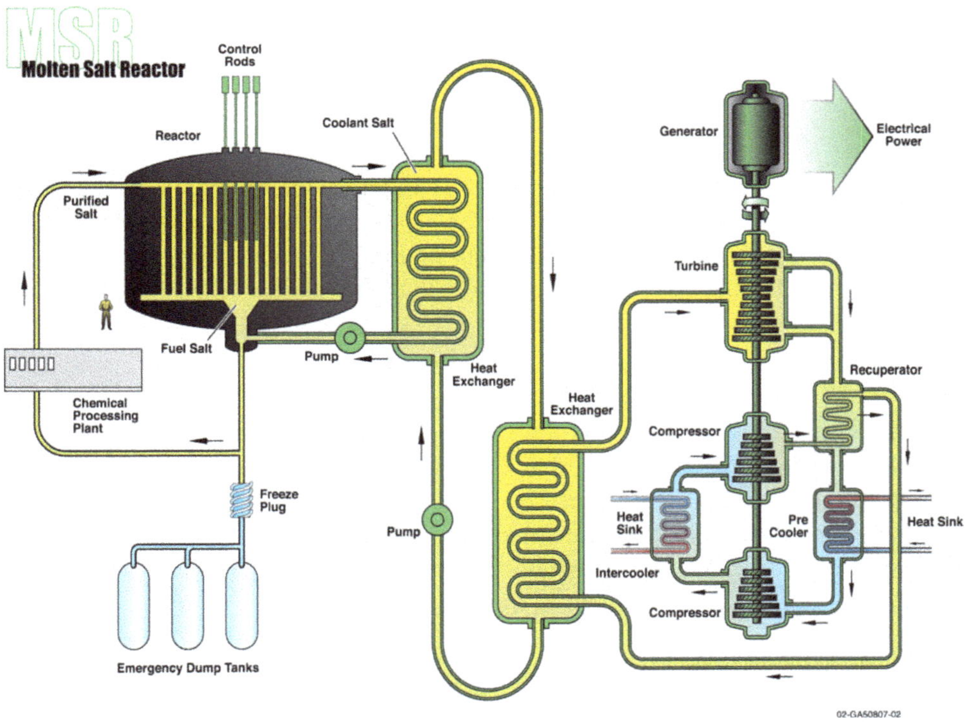

Figure 6.2. Reference design for the MSR.

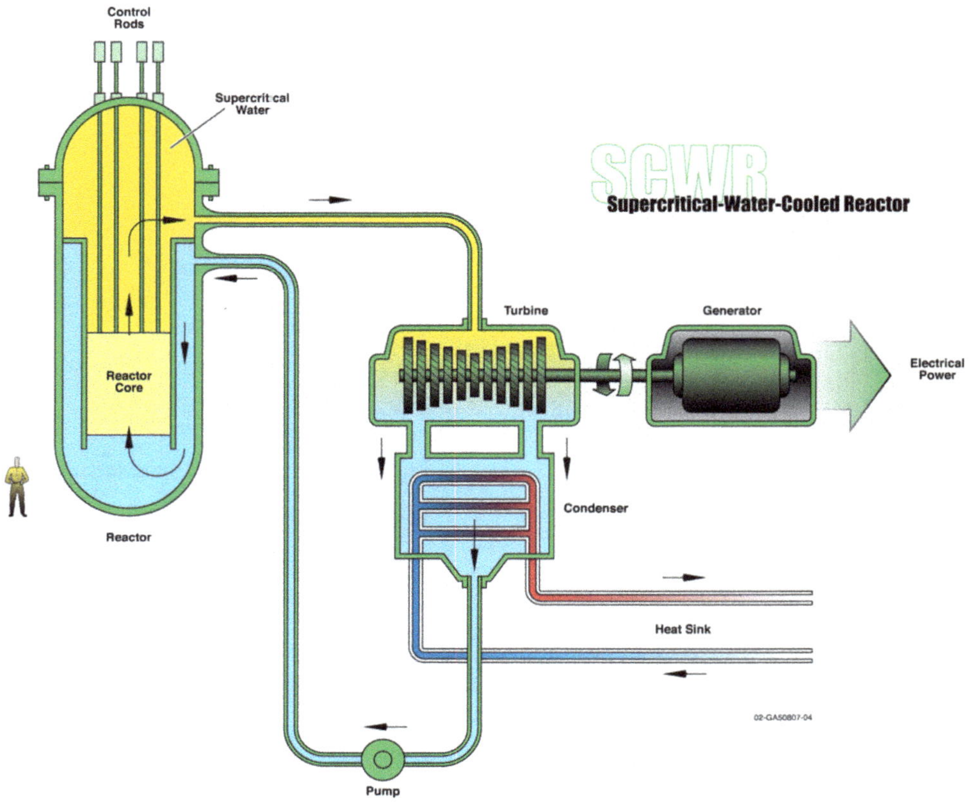

Figure 6.3. Reference design for the SCWR.

temperature above which the fluid becomes a vapour under any pressure. In the case of H_2O, the critical point is 647 K, as above this temperature only vapour exists. So, in this design liquid water enters the system and supercritical steam exits. The key criteria for the design are shown in table 6.1 and the schematic design is shown in figure 6.3.

6.2.4 Gas-cooled fast reactor (GFR)

This design works in a similar manner to the VHTR, but whereas the VHTR uses moderation of neutrons in increased fission, this design formally does not have extra moderation other than that provided by the fuel. The design goal of this design is to transmute undesirable and fertile isotopes into usable fuel, in a manner that is more efficient that current reactors. The key criteria for the design are shown in table 6.1 and the schematic design is shown in figure 6.4.

6.2.5 Sodium-cooled fast reactor (SFR)

This proposed design takes advantage of fast neutrons, i.e. increased transmutation, to develop a reactor technology that could be used to reduce the inventory of

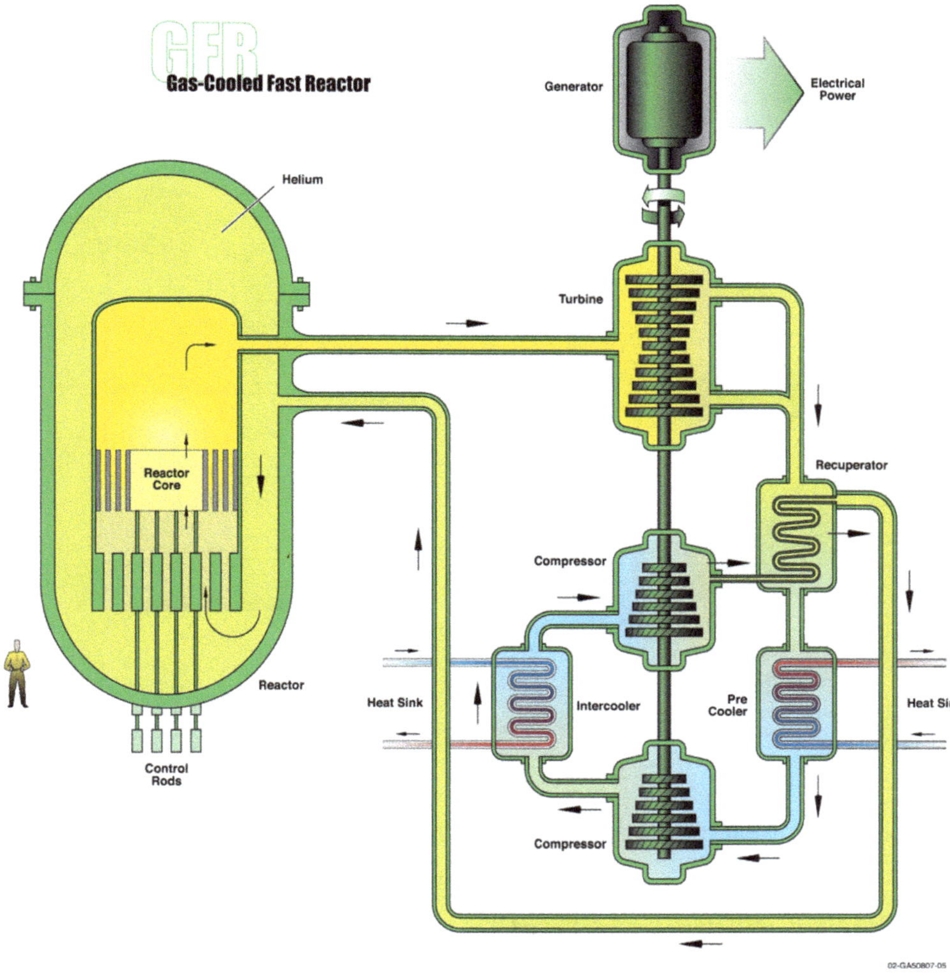

Figure 6.4. Reference design for the GFR.

high-level waste. Using liquid sodium as coolant allows for reasonably high temperatures of output, with a lower risk of boiling, i.e. Na melts at 271 K and boils at 1155 K. This reactor design builds upon the experiences of the Prototype Fast Reactor (PFR) at Dounreay in the UK, and the USS Seawolf (see chapter 5). The key criteria for the design are shown in table 6.1 and the schematic design is shown in figure 6.5.

6.2.6 Lead-cooled fast reactor (LFR)

Whereas the US (and the UK, for that matter) navy used small PWR for power, the Soviet/Russian navy used liquid lead-cooled reactors, particularly in the Alfa class of submarines. This technology works in a similar manner to the SFR and MSR designs. Its primary focus is as a breeding reactor, operating in a manner similar to the SFR, but it is one that can be built in a factory with a closed-fuel replaceable fuel

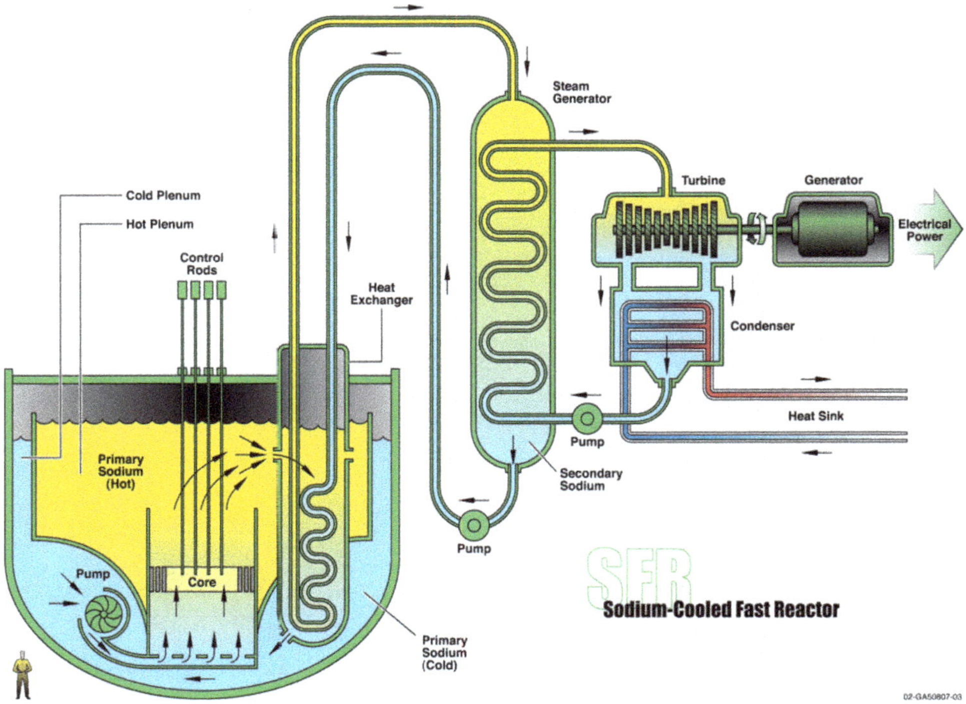

Figure 6.5. Reference design for the SFR.

core. The key criteria for the design are shown in table 6.1 and the schematic design is shown in figure 6.6.

6.3 Material challenges in GenIV

As most material scientists will tell you, by their very names the six GenIV designs imply significant challenges for the materials being used. To use the VHTR as an example, running the high temperatures expected requires a material that retains shape and strength, while also allowing neutron transport. Rather than examining each design in detail and discussing the challenges associated with it, we shall examine the key requirements across all, and contrast the requirements for each based on common topics.

6.3.1 Fuel

In many of the GenIV designs the nature of the fuel needs to be reconsidered. Often this can be due to the nature of the core environment, whilst in others there is such a fundamental change that current fuel is no longer usable. As can be seen in table 6.1, the majority of the fuel types are solid, so in principle we could use UO_2, and in some cases that has been proposed. However, if you remember back to where we discussed fuel, there are issues with the long-term use of UO_2, hence in some cases, the LFR for example, metallic fuels have been proposed.

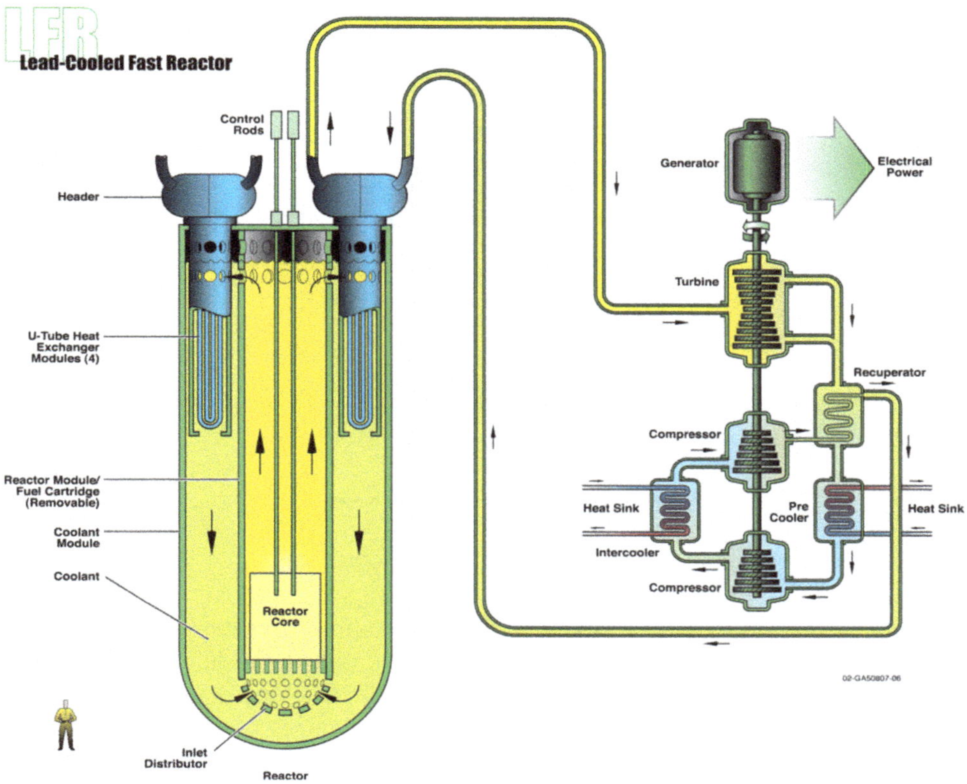

Figure 6.6. Reference design for the LFR.

For use within the VHTR and GFR, which have much commonality in their design, the fuel runs at much higher temperatures than are currently experienced in most power reactors. The proposed fuel design for these reactors is based on the TRISO methodology, which takes the concepts of fuel and cladding and applies them in a unique way.

6.3.2 TRI-ISOtropic (TRISO) fuel

Previously, when we have discussed fuel and cladding we have considered them separately and then in combination. Since they are two separate materials that are used together, this is okay. What if we were to change the design so that the cladding and the fuel are now considered as one?

This methodology is how TRISO fuel is designed to work [2]. TRISO particles are ~1 mm in diameter and work in a manner similar to an onion, with a kernel of UO_2 in the centre of the sphere and concentric layers coating the UO_2. These layers are based on carbon for heat conduction and material compatibility, and SiC for material strength. A schematic of a TRISO particle is shown in figure 6.7. The concept for TRISO was developed during the DRAGON programme and allows much higher temperatures to be used. The key advantages that TRISO brings are: the internal moderation of neutrons (using carbon), negating an external moderator;

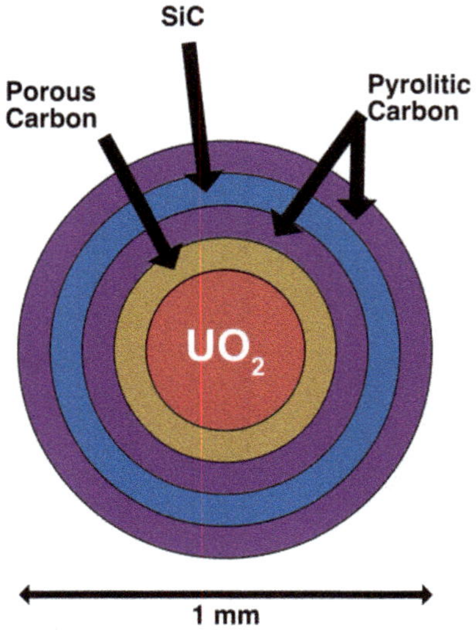

Figure 6.7. Schematic of a TRISO particle.

the ability to retain fission products; and the limitation of failures to one particle. Unfortunately, in light water reactors TRISO is non-ideal and has not been used.

There is one potential problem with using TRISO that has generally been overcome, but gives an insight into the material challenges that impact on GenIV development. This challenge arises from differential thermal expansion, which has been accommodated through the use of graphite-based layers. As can be seen in figure 6.7, TRISO contains concentric layers of UO_2, C and SiC, and when undergoing thermal expansion within the core, it would be expected that the differential rate effects of thermal expansion, and the issues it gives rise to, such as stress/strain, could cause problems. This is further complicated when one considers the crystal structures of the components. In chapter 3, we discussed the cubic nature of UO_2 and how it behaves isotropically, whereas hexagonal Zr-based materials behave aniostropically. In TRISO, since we are using UO_2 as the kernel, the expansion would therefore be isotropic and predictable. In contact with the UO_2 are layers of carbon, adopting the graphite polymorph and hence hexagonal, thus it would be expected to behave anisotropically. Surrounding these carbon layers is a layer of SiC, which is where problems can arise, which relate to the structure of SiC. There are multiple polymorphs of SiC, with cubic and hexagonal being the two extremes. So, which polymorph would be better for use within TRISO?

If it was being based purely on thermal expansion, it could be argued that a cubic system is better, as the expansion is isotropic. But, in TRISO the SiC layer is contained between two carbon layers, which are hexagonal in nature, providing extra barrier, so is cubic still better?

This is a complex question to answer, as there are benefits to both cubic and hexagonal lattices. However, examination of sample TRISO particles has shown that the cubic polymorph is formed during the coating of the pellets [3]. How does the particle behave when used as part of a fuel?

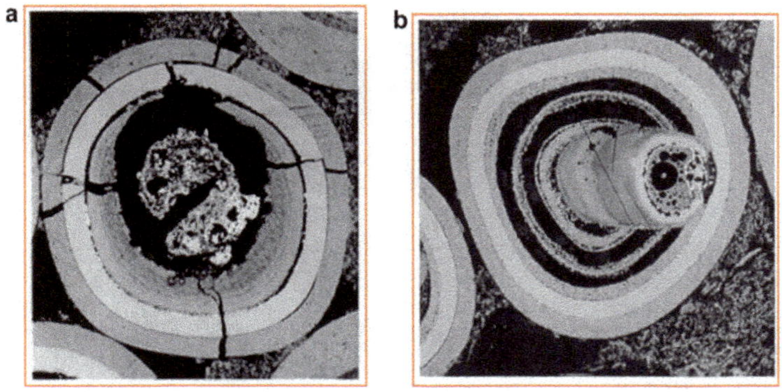

Figure 6.8. Examples of TRISO particle failure [4].

This is where the nature of the TRISO particle comes into its own. If we consider the hardness of the three material types, and in particular the nature of the carbon used, certain behavioural trends can be expected. Graphite is the softest of the three materials present in TRISO, after all it is used in pencils, and shrinks when irradiated within a nuclear reactor [3], which can give rise to potential problems when used within cores. Therefore, if we place graphite between two expanding materials, it could be expected to accommodate some of the expansion by compressing. So, how do the layers behave and why are they there?

As can be seen in figure 6.7, the central core of (mostly) UO_2 is surrounded by a layer of porous carbon, which is then surrounded by a layer of pyrolytic carbon, a more dense layer, which is in turn surrounded by SiC, with an outer coating of pyrolytic carbon. So, which layer does what?

In simple terms, the UO_2 provides the fuel for fission, the porous carbon layer absorbs the fission products/gases and some of the induced damage, the pyrolytic carbon absorbs more damage and the SiC layer provides the final barrier to fission products. The outer layer of pyrolytic carbon provides strength, and as carbon shrinks when undergoing radiation damage, provides compressive strength, helping to keep the particle intact.

Generally, failure within TRISO particles is local and does not cause major issues with fuel/reactor operation. Some examples of failure are shown in figure 6.8.

There is an alternative to TRISO that contains an extra layer, this time a burn-up absorber such as Eu, which helps to balance the operation of the fuel. This layer is placed between the UO_2 and the first carbon layer, preventing neutrons from hitting the UO_2 and initiating fission. As the layer is burnt up, the UO_2 can then fission, allowing a more balanced load and burn-up across the core.

Each TRISO particle is further embedded into larger arrangements, forming larger balls or pebbles, or within hexagonal-shaped blocks formed from graphite/ZrC. This is the designed fuel for the VHTR and GFR reactors, and shown in figure 6.9.

6.4 Containment

In some designs, such as the SCWR, the nature of the fuel does not need to change, it is the method of containment that must. Before we discuss the modifications in

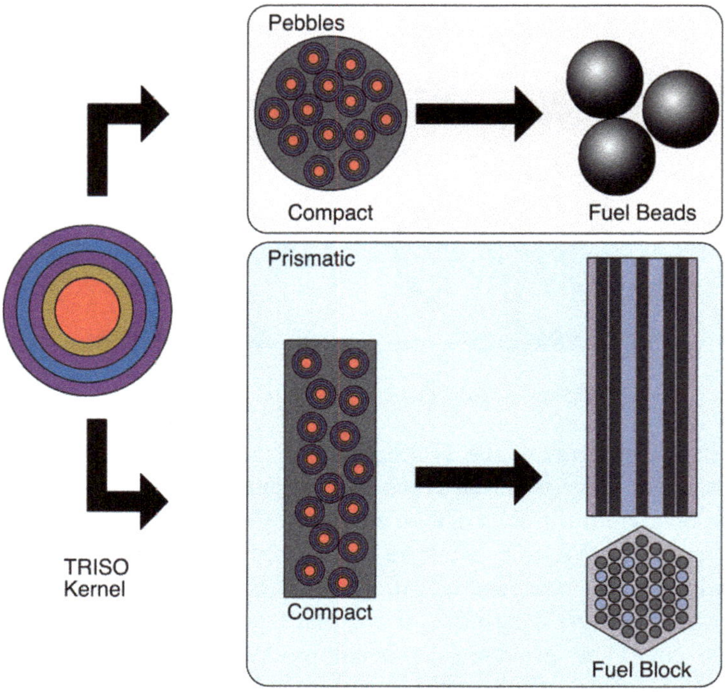

Figure 6.9. Schematics of prismatic fuel blocks and larger beads of fuel.

cladding that are required, we should remind ourselves of what cladding is designed to provide.

6.4.1 Key requirements for cladding

In chapters 3 and 4 we discussed the key requirements for clad, and how they behave in a light water reactor. The key requirements for cladding are, therefore:

(i) melting point—must not be too low, to allow for efficient reactor operation;
(ii) chemical compatibility—must not be a highly reactive element, i.e. must be chemically inert with both the fuel and coolant used;
(iii) manufacturable—can it be formed into the required shape/geometry;
(iv) mechanically stable—it is required to contain fuel, so therefore should not undergo major changes within the reactor core (later we will discuss some of the causes of cladding changes);
(v) acceptable neutron absorption—while it is ideal to have the lowest possible, it may not be practicable, for example the lowest neutron absorption for any naturally occurring element is seen in oxygen, and we obviously cannot make cladding out of that.

When identifying the most important criteria for cladding in GenIV reactors, we need to understand more fully the conditions being used. For example, in the LFR design the coolant is liquid Pb, whereas in the SCWR the coolant is highly energetic water vapour. It can therefore be argued that the two key criteria here are the melting point and chemical compatibility, and in some cases the chemical compatibility of the cladding with the coolant is the key criterion. Therefore, using this methodology, what cladding would be best for the SCWR design?

In reactors such as the PWR- and BWR-based designs, the cladding is Zr-based, which has run safely for thousands of operational hours. However, there are have been two significant cases where it has operated less successfully: at Three Mile Island in the USA and at Fukushima in Japan. Whilst Zr-based cladding can be used successfully when the coolant is water, when steam is introduced the corrosion escalates rapidly, as seen at both Three Mile Island and Fukushima. Taking the supercritical nature of H_2O in the SCWR, i.e. vapour, it should be apparent that Zr-based cladding is no longer viable. Alternatives such as Ni-based and Fe-based alloys have been proposed for use, with the general acceptance that chemical compatibility trumps neutron absorption, i.e. the cladding has to survive long enough for the core to operate. The gains obtained from the increased thermodynamic efficiency will help to offset the loss in neutron transport.

Similar considerations will have to be made for the SFR, LFR and MSR designs, as the corrosive nature of the coolant needs to be accounted for and accommodated.

6.4.2 Liquid fuel

So far we have discussed solid fuel, e.g. U, UO_2 and MOX, and briefly mentioned the potential for liquid fuel. When using a molten salt as the coolant, the option exists to use both solid and liquid fuel, and when using liquid fuel, it is based within the coolant, as we saw in chapter 5. Liquid fuel has many advantages over solid fuel, and by far the greatest of these is the ability it gives to remove the fission products through purification and return the unused fuel back to the core. Whilst this combined fuel/coolant may be ideal from a neutronic/reactor operation viewpoint, from a materials perspective we now have the combination of enhanced radiation damage and corrosion. So why are these different, and more challenging, than when using a solid fuel and a coolant such as H_2O?

Firstly, let us discuss the nature of the molten salts proposed for use within a reactor, i.e. based on fluoride. Generally, these salts, as most chemists will tell you, are ionic in nature, and have multiple uses. One important application of molten salts is their ability to dissolve other compounds/elements, for example, Na_3AlF_6 (cryolite) is the host for the extraction of aluminium from aluminium oxide using electrolysis. In this example, the Al_2O_3 dissolves into the Na_3AlF_6, highlighting the ability of molten salts to dissolve other systems, indicating a potential challenge within a nuclear reactor. Is it possible to use a metric to predict the likely dissolution or reaction behaviour of elements with each other?

If we use the electrochemical potentials for the elements, then a predicted reactivity series can be made, as shown in figure 6.10.

If a purely thermodynamic approach is made to the likelihood of formation for a metal halide from its constituent elements, and the Gibbs free energy for formation is used as a metric, we can obtain a ranking scheme, as shown in figure 6.11.

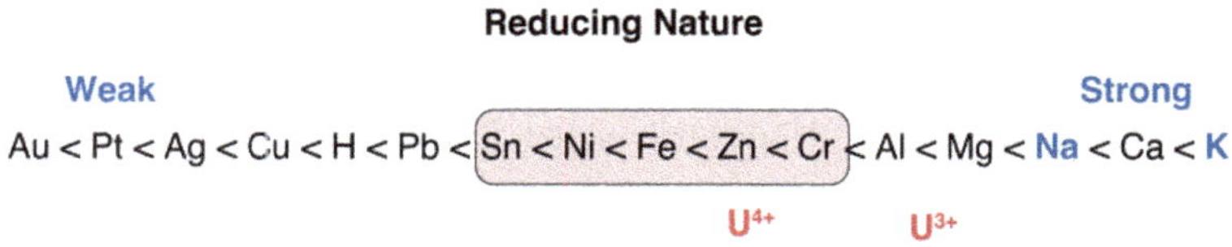

Figure 6.10. Predicted reducing behaviour of the elements from the strongly to the weakly reducing; U^{3+} and U^{4+} have been added, with their predicted positions shown.

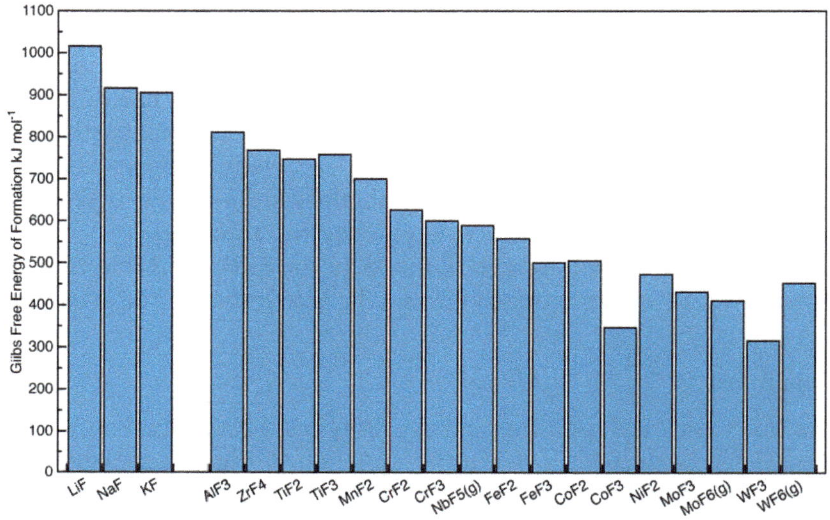

Figure 6.11. Predicted Gibbs free energy for formation of metal halides, based on data presented in [5].

Table 6.2. Major components in alloys proposed for use within molten salt reactors [6, 7].

Alloy	Major constituents (>1%)
Hastelloy-N	Nickel, molybdenum, chromium, iron and silicon
Hastelloy-X	Nickel, chromium, iron, molybdenum, and cobalt
Inconel-617	Nickel, chromium, cobalt, molybdenum, and aluminium
Haynes-230	Nickel, chromium, tungsten, molybdenum, cobalt and iron

As can hopefully be seen in figures 6.10 and 6.11, Ni,Fe, Cr and Zn would be expected to be stable to molten salt attack. Indeed, many of the proposed alloys for use within an MSR are based on Ni and Cr, as shown in table 6.2.

If the alloy components in table 6.2 are compared with the rankings in figures 6.10 and 6.11, one would expect the alloys to remain as alloys, as they are less thermodynamically likely to form over NaF, KF and LiF, the major components within the melt. So, if thermodynamically they are unlikely to react, why are we concerned?

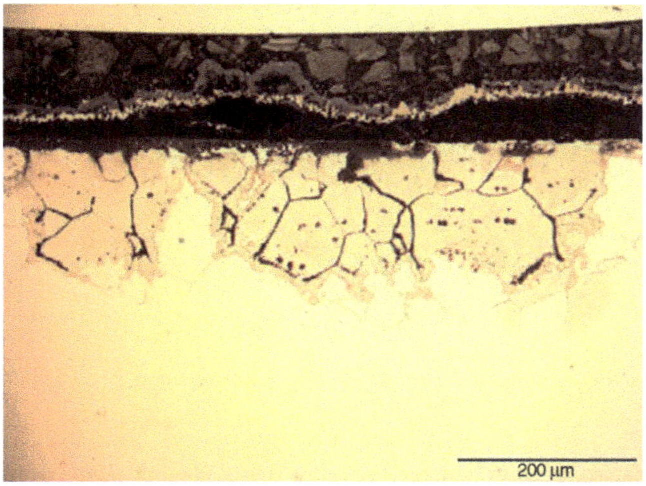

Figure 6.12. Corrosion of a Ni-based alloy from Te present in the molten salt [8].

If we were only considering chemical corrosion of the starting materials, the concern for long-term stability would not be as great as it is, as it could be modelled using standard thermodynamics and reaction kinetics. However, that would overlook one important issue that arises when materials are used within a reactor core and contain nuclear fuel—the role that fission has to play.

As we discussed in previous chapters, once a fission chain reaction begins, we longer have the starting composition, with multiple potential fission products that can arise. Now the situation is more complex, and it is how these fission products behave that affects the corrosion dramatically. For example, the common fission products of Te, I, Mo, Tc, Cs, Sr and the lanthanides can now form compounds within the salt, for example, the reaction between Ni and Te, and shown in figure 6.12:

$$Ni + Te \rightarrow Ni_xTe_y. \tag{6.1}$$

If we take two specific examples of fission product formation and see how they would behave, we gain an insight into the material challenges that need to be overcome. The first of these is the formation of free F within the molten salt, arising from the fission of UF_4:

$$UF_4 + n \rightarrow LnF_3 + M + \frac{1}{2}F_{2(g)}. \tag{6.2}$$

This free F is highly reactive, increasing the corrosive nature of the system.

The second concern is more insidious and involves a reaction between the UF_4 in the melt and the Cr present in the alloy. As was shown in figure 6.10, Cr sits between U^{4+} and U^{3+}, so we can assume it is more reactive than U^{4+} and less reactive than U^{3+}, therefore what can we expect to happen?

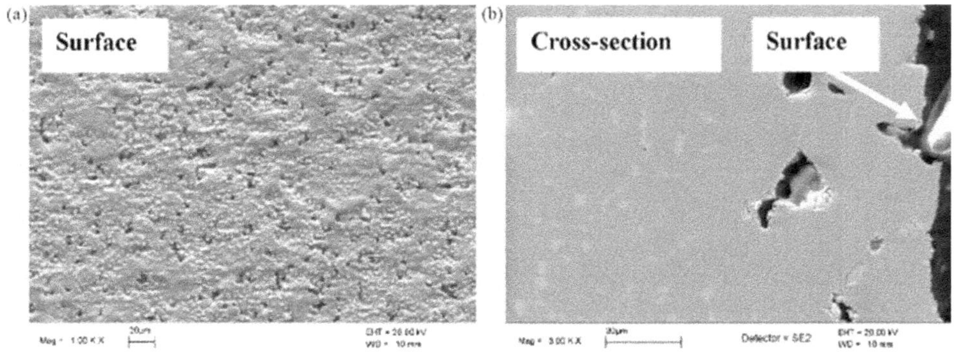

Figure 6.13. Corrosion of Hastelloy-N after 500 h in molten salt at 1125 K [5].

Effectively, Cr reduces U^{4+} to U^{3+}, giving rising to enhanced corrosion, as shown in figure 6.13:

$$2UF_4 + Cr \rightarrow 2UF_3 + CrF_2 \tag{6.3}$$

So why do we still use Cr, if its presence increases corrosion? In many cases, having Cr present makes it easier to manufacture the components being used and helps to improve their quality. However, that is not to say that further research is not being undertaken to develop alloys that are Cr-free, indeed that is the case and this is one of the areas being developed for MSR.

As with the coating of cladding to prevent LOCA (chapter 4), coating MSR materials with e.g. carbides/nitrides to enhance their durability is also a possibility.

6.5 Radiation damage

So far in this chapter we have looked at the impact of GenIV designs on fuel and cladding/containment, but there is one area that we have overlooked that impacts dramatically on the range of potential materials available for GenIV technologies.

As part of the design process for the GenIV reactor family, extended operational life and higher burn-up of nuclear fuel are key goals. It would be expected that the materials within the reactor would experience higher levels of induced damage. Indeed, the expectation in the MSR design is for a level of damage approaching 150 dpa [1]. If the effects of damage are coupled with the expected temperature range within a core, the issue we discussed in chapters 2 and 5, this now becomes highly problematic. For example, how a material behaves when being damaged by neutrons at 1200 K can be dramatic, and could quickly lead to component failure. How can a system be designed that takes into account the effects of chemical corrosion and radiation damage? This is one of the key questions that need to be answered, and it is a particularly apt one for the MSR design.

When considering the effects of radiation damage in reactor materials outside of the fuel, the damage mainly arises from neutron impact, as we discussed in chapter 2. The pressure vessel and those parts, which are not generally easily replaceable, do

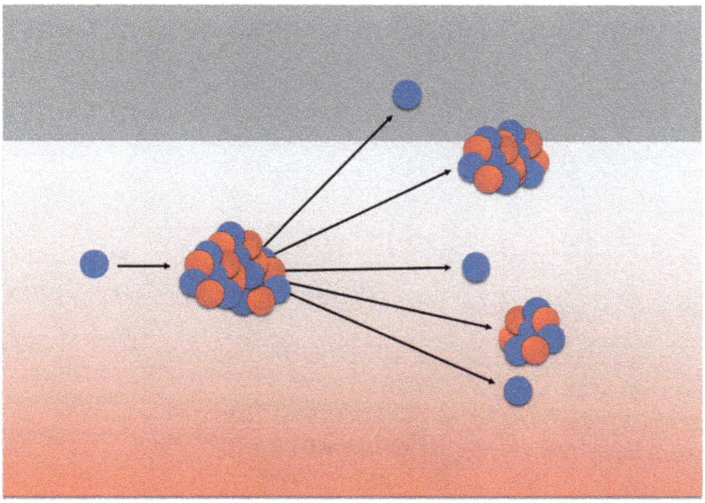

Figure 6.14. A fission event occurring that impacts on the containment of the molten salt.

not experience very high levels, because of their distance from the source. In fuel, the damage arises from multiple sources, neutron impact, fission and fragment production, and radioactive decay. This form of damage can interact with the cladding present, modifying its properties, but this is replaced when the fuel is removed and is usually not load bearing within the core. Now consider a reactor design where the fuel is in direct contact with the containment, and where radiation damage can arise from this—what could happen?

To answer this we need to remind ourselves of the outcome from fission, other than heat, and generally we would expect two or more fission fragments and some neutrons. Each of these particles will have a momentum that interacts with its local surroundings. Now consider these two scenarios: firstly, the fission happening in the centre of the fuel, away from containment, and secondly, an event in close proximity to the containment. What would be the outcome in each?

In the first scenario, the containment is likely to be unaffected, as the molten salt will absorb most of the particle momentum, inducing further fission or dissolving the fission products. It is the second scenario that induces further damage, from the fission fragment and neutrons impacting on the material, as shown in figure 6.14.

These events will cause radiation damage and induce cracking, stress/strain, volume expansion, etc, while they can also induce gas bubble formation, leading to more points of weakness in the material. Each of these outcomes is deleterious to the metallic structure, and we have overlooked the possibility of transmutation potentially modifying the composition of the alloy and inducing He bubble formation:

$$^{58}_{28}\mathrm{Ni} + ^{1}_{0}\mathrm{n} \rightarrow ^{54}_{26}\mathrm{Fe} + ^{4}_{2}\mathrm{He}. \quad (6.4)$$

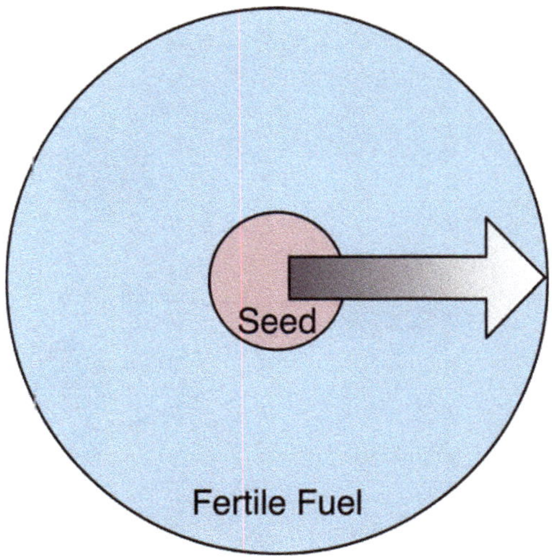

Figure 6.15. Animation of the travelling wave process within a reactor core.

6.6 Alternative reactor technology

So far we have considered the historical GenI–GenIII/III+ and the advanced GenIV reactor designs, but these are not the only designs being put forward. There are three other types of technology that suggest material challenges, and which can work in parallel with GenIV designs, if successful. The first one is the development of an interesting proposal for fission and can use fertile rather than fissile-based fuel.

6.7 Travelling wave reactor

Imagine, if you will, a trail of gunpowder that when lit burns from one end to the other, similar to Wile E Coyote lighting gunpowder when trying catch the Road Runner. Now replace the gunpowder with fertile nuclear fuel, and the initial flame with fissile material, and in essence that is how a travelling wave reactor would work. Currently, fuel in reactors tends to be moved around in the core, and fresh fuel replaces older fuel [9]. The travelling wave reactor does not need this, as once a core is started it can run until completion. So how would this work in a core?

Rather than burning from one end to the other, this would burn from the centre of the core out, in a manner similar to figure 6.15.

The specific key requirements for the travelling wave design are fertile material, e.g. ^{238}U or ^{232}Th, and an initial seed of fissile material, such as ^{239}Pu or ^{235}U. The other requirements are, as with the majority of the GenIV designs, materials that are capable of withstanding high levels of radiation damage. The key difference between this design and other reactor designs is that once this is switched on, it is intended to run continuously until the fuel has been burnt, with minimal interaction. This could be 40–50 years, an intriguing prospect for reactor and materials design.

This design is being used as the basis for the reactor technology being developed by Terrapower LLC [10].

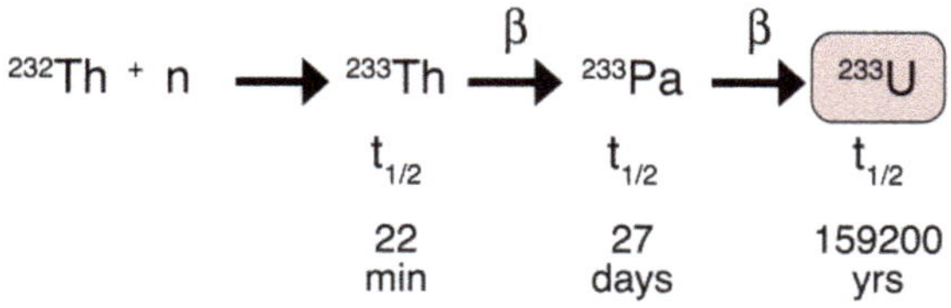

Figure 6.16. Simplified transmutation process from ^{232}Th to ^{233}U.

6.8 Thorium reactors

So far, when discussing fuel we have predominantly focused on fissile materials, such as ^{235}U and ^{239}Pu, but what happens if we base a fuel cycle on the use of fertile materials, such as ^{232}Th or ^{238}U? In fact, in the early reactors ^{238}U was used to breed ^{239}Pu, but what about using ^{232}Th, does it have any advantages over $^{235}U/^{238}U$?

6.8.1 Properties of thorium

Thorium naturally has six isotopes, of which ^{232}Th is the most prevalent, having a half-life of 1.405×10^{10} years, and it is formally considered metastable as it undergoes alpha-decay, forming part of the thorium decay chain. The composition of the Earth's crust contains about four times more thorium than uranium, and while they are often found together, thorium can be found in other locations.

One specific area where Th has an advantage over U is that Th^{4+} is the stable oxidation state, whereas for U, multiple states exist, as discussed in chapter 3. This arises from the outer electron configuration being *[Rn] 7s² 6d²*, which when oxidised becomes *[Rn]*. This single stable oxidation state gives rise to ThO_2 being more chemically stable than UO_2 under oxidising conditions. However, as outlined previously, ^{232}Th is fertile and not fissile, and needs to be transmuted to a fissile isotope before it can be used as fuel. How can this be achieved and used within a reactor core?

6.8.2 Thorium-based fuel

Before we discuss how ^{232}Th-based fuel is used within the core, we should understand how the process of transmutation occurs, going from ^{232}Th to ^{233}U, which is the fissile isotope. The process is both trivial and complex, impacting on how a reactor can operate.

Transmutation of ^{232}Th to ^{233}U

If we consider this from a purely mechanistic approach, the transmutation from ^{232}Th to ^{233}U is ideally three steps, requiring one neutron and two beta decays (figure 6.16). The process, once initiated, will take approximately 27 days to complete. However, is it that easy or simple?

In reality, the process can be more complex, giving rise to more products, as shown in figure 6.17. This more complex process highlights one of the issues of using ^{232}Th as a source of fertile fuel, which concerns the formation of protactinium, in

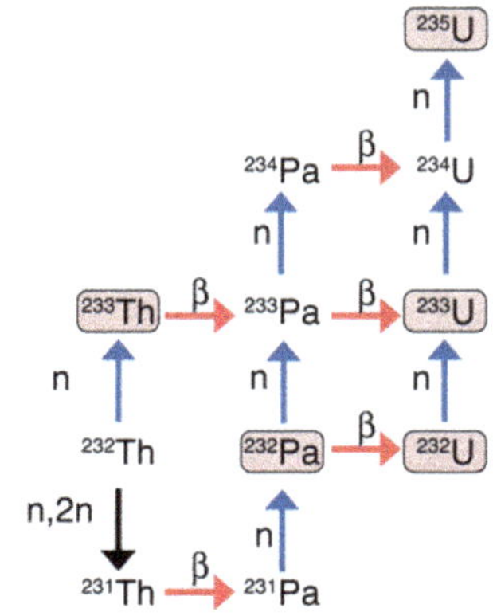

Figure 6.17. Transmutation process starting with ^{232}Th.

particular, ^{231}Pa and ^{233}Pa. These isotopes are problematic for two reasons: ^{231}Pa has a half-life of ~3.3 × 10^4 years, increasing the radiotoxicity, and has a large neutron absorption of ~200 barns, whilst ^{233}Pa has a shorter half-life of 27 days, but again has an appreciable absorption of ~40 barns. Each of these isotopes can transmute further before undergoing decay through neutron absorption, giving rise to different transmutation products, for example, ^{232}U.

So far, we have discussed the outcome of transmutation of ^{232}Th, but where can we obtain neutrons?

How to induce transmutation

Unfortunately, it is not easy to obtain neutrons for transmutation, free neutrons have a half-life of ~10 min and so cannot be stored, but luckily there are two major sources we can use. The first source is purely material undergoing fission, e.g. actinides, using these we could surround the ^{232}Th with ^{239}Pu/^{235}U and other actinide isotopes as an initiator for the reaction. Once started, the reaction should be self-sufficient in providing neutrons.

As was proposed in the travelling wave reactor design, using such a method could lead to a reactor that can run for long periods of time without refueling. However, this may not be an ideal approach. An alternative approach would be to generate a beam of neutrons using a proton accelerator irradiated into a target, directing it into a sub-critical volume of material and initiating transmutation and fission. This approach has the advantage that the reactor could be expected to switch off more easily, but how would the reactor work? A similar process occurs at the ISIS facility at the Rutherford Appleton laboratory in the UK, which uses a proton beam into a target to spallate neutrons, which are then used for analysis materials.

As can be seen in figure 6.18, the reactor core itself would be irradiated by a source of neutrons and used to initiate transmutation and fission. This process is challenging in terms of materials, as if you think back to previous chapters, we are now inducing damage in materials directly, and thus need to develop materials that will resist such damage. What are the options available to minimise the damage effects within the materials?

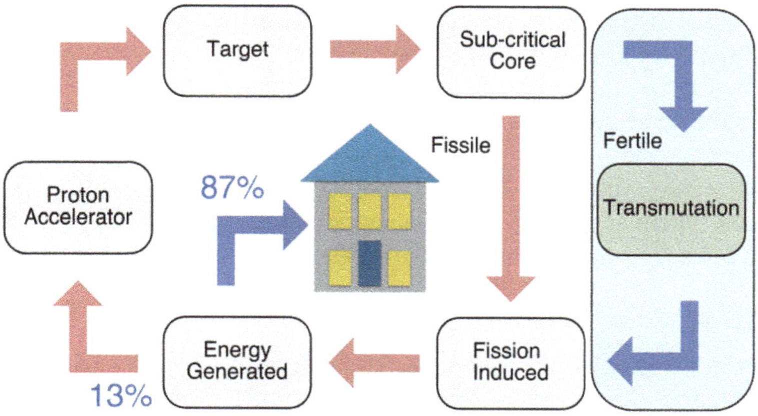

Figure 6.18. Schematic of the accelerator-driven method for reactor operation.

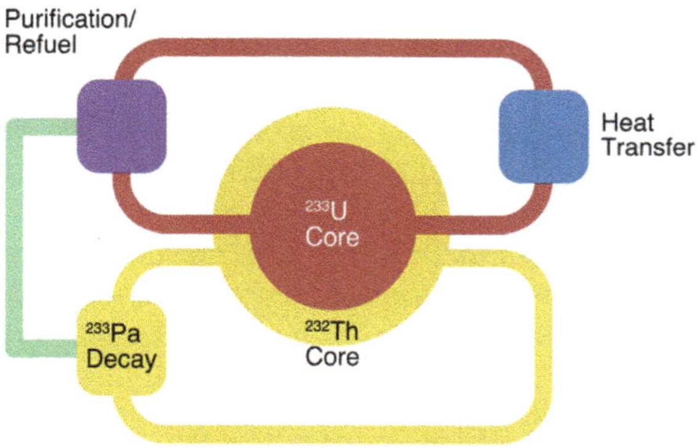

Figure 6.19. Schematic of the liquid fluoride thorium reactor.

6.8.3 Thorium reactor designs

As with fissile-based fuels, multiple options have been put forward for Th-based reactors, using a range of fuel options and various methods for generating fissile material. Rather than go through each, we shall focus on one specific design, which has commonality with the GenIV MSR design we discussed earlier, i.e. one using a liquid fuel and liquid coolant in combination.

Liquid fluoride thorium reactor (LFTR)

This design takes as its basis the use of neutrons from fission for the initiating step in transmutations. In this design there are two molten salt loops, which couple. Firstly, the primary circuit, similarly to the MSR GenIV design, contains the fissile material, in this case ^{233}U, undergoing fission and generating heat and neutrons. This circuit then passes through purification, where the new fissile material is added, shown in figure 6.19. But how do we generate this new fissile material?

Contained around the fission core is a blanket of molten $^{232}ThF_4$ salt, which circulates and undergoes a two-step process. Firstly, the molten salt is left to settle for a time, allowing isotopes such as ^{233}Pa to decay, and then it is purified with the used fissile material and introduced into the core.

This reactor has the potential to run for extended periods of time, but still experiences the same problems that the MSR GenIV design faces, such as radiation damage and corrosion.

So far we have discussed large reactors that would be used to power large communities, such as cities. There is an alternative approach that focuses on generating electricity near to where it is needed: small modular reactors.

6.9 Small modular reactors (SMR)

These reactors are based on the premise that large reactors, such as the ABWR, EPR and AP1000 family, may not be ideal for some applications. Taking the economics of operation into account, in countries where there is a mature power grid, reactors outputting ~1GWe can contribute power from dispersed locations. What happens if the grid cannot accommodate such power levels, or if the application is remote, such as an isolated community in the deserts of western Australia? This is the idea behind small modular reactors (SMR).

The IAEA classifies SMR as reactors designed to output less than 300 MWe, and there are multiple designs that match this criterion [11]. What are the material challenges for SMR development?

In essence, they are the same as those for larger reactor development, the ability for the reactor to run for extended periods of time with minimal problems occurring. The key challenges of radiation damage, high temperature materials behaviour and the effects of corrosion are generally the same, but within an SMR there are also differences.

For example, how would you refuel a reactor located in the middle of the western Australian desert, where the only source of energy is the reactor itself? In a large-scale reactor, this process could take more than a few weeks, so how would it work in a SMR? The design of many SMR is based on the replacement of the core, the exchange of the reactor itself, or the ability to run the reactor indefinitely. For example, using the Flexblue design as a example of the thinking behind SMR, this reactor is placed at the bottom of the sea, near to where the power is needed, and can run for up to 60 years, with refuelling every 40 months [11]. The refuelling is where the SMR rationale comes into its own, rather than talking the reactor offline for extended periods, the unit can instead taken back to its support facility and refuelled. In the meantime, a second module can be connected in its place, minimising downtime.

In essence, the SMR means taking each of the options we have discussed so far in this chapter for GenIV and alternative reactor designs and shrinking them down. So, what are the designs being proposed?

6.9.1 Proposed designs

As you can probably imagine, as with their larger cousins, there are multiple design concepts put forward for SMR, but they can be classified using the same criteria and broken down into the same broad families shown in table 6.3.

Table 6.3. Examples of proposed SMR designs [11].

Reactor Type	Coolant	Examples
PWR	H_2O	Flexblue, NuScale
BWR	H_2O	DMS
AHWR	D_2O	AHWR300-LEU
Gas cooled	He	GT-HTR300, PBMR
Sodium cooled	Na	4S, Prism
Lead cooled	Pb, Pb/Bi	Brest-OD-30, G4M

In many cases, an SMR design has evolved from a larger design and been reduced in size, whilst some others have evolved from reactors in other applications. For example, the navies of the USA, Russia, the UK and France have all had experience of running nuclear reactors onboard naval vessels. Indeed, the Russian navy has gone further and powered civilian icebreaker vessels with nuclear reactors to great effect.

As a result of the commonality with their larger cousins, the majority of the material challenges and developments run in parallel with GenIV development. However, one key difference is that rather than a small number of large-scale reactors being built, a large number of smaller reactors would be built, taking advantage of the economies of scale large-scale manufacturing can bring.

In this chapter, we have examined the challenges and goals of the GenIV programme and alternative reactor technologies, and discussed the development of small modular reactors. If we contrast this chapter with chapter 5, we can see that where small-scale reactors developed for naval/aircraft applications became the basis for many large commercial reactors, we have now returned to developing small-scale nuclear reactors.

So far, we have only considered nuclear materials and their use within reactor/fuel development, and we should now consider the material challenges within the disposal of the radioactive waste formed within a reactor. This is the focus of the next chapter.

References

[1] GenIV International Forum 2002 *A technology roadmap for generation IV nuclear energy systems GIF-002-00*

[2] Price M S T 2012 The Dragon Project origins, achievements and legacies *Nucl. Eng. Des.* **251** 60–8

[3] IAEA *High Temperature Gas Cooled Reactor Fuels and Materials IAEA-TECDOC-1645*

[4] Olander D 2009 Nuclear fuels—present and future *J. Nucl. Mater.* **389** 1–22

[5] Olson L C, Ambrosek J W, Sridharan K, Anderson M H and Allen T R 2009 Materials corrosion in molten LiF–NaF–KF salt *J. Nucl. Mater.* **130** 67–73

[6] Haynes International *High Temperature Alloys*

[7] Special Metals Corporation

[8] Delpech S, Cabet C, Slim C and Picard G S 2010 Molten fluorides for nuclear applications *Mater. Today* **13** 34–41

[9] Feinberg S M 1958 Discussion comment *Rec. of Proc. Session B-10, ICPUAE (United Nations, Geneva, Switzerland)*
[10] Ahlfield C *et al* 2011 Conceptual Design of a 500 MWe traveling wave demonstration reactor plant *Proc. ICAPP 2011* paper 11199
[11] IAEA 2012 *Status of Small and Medium Reactor Designs* (IAEA)
IAEA 2014 *Advances in Small Modular Reactor Technology Developments* (IAEA)

IOP Publishing

Nuclear Materials Science

Karl Whittle

Chapter 7

The challenges of nuclear waste

Up till now, we have discussed nuclear materials from the viewpoint of reactor technology, but there is also the equally challenging problem of safely disposing of nuclear waste, once formed. This chapter will outline the major challenges and how they relate to the development of waste forms, but will not go into great detail about topics such as geological disposal, specific methods for containment of nuclear waste, etc. These are questions that could form the basis of a book in themselves, and in this chapter we shall discuss the issues more generally and see how they relate to the topics that we have already discussed.

7.1 Sources of nuclear waste

We need to define nuclear waste more accurately before we can understand what its challenges are. Using the IAEA guidelines [1] as the metric by which we classify nuclear waste, there are six types, going from 'exempt' through to 'high level'. The classifications are defined as follows.

(i) Exempt: material that is exempt from control for radiation and environmental protection purposes.
(ii) Very short lived (VSLW): decays over a few years to a level that becomes exempt (can be managed by delay-decay).
(iii) Very low level (VLLW): suitable for regulated landfill under control, due to the concentration and activity, or with a small long-lived inventory.
(iv) Low level (LLW): disposal in an engineered facility with assured containment for 100 y, long-lived radio nuclide inventory.
(v) Intermediate level (ILW): disposal in an engineered subsurface facility (100 m), with containment exceeding timescale of controls. Heat dissipation is not required.
(vi) High level (HLW): inventory requires disposal in a GDF (1000 m) and heat mitigation mechanisms are required (or at least considered).

doi:10.1088/978-0-7503-1104-5ch7

Since the majority of material challenges in immobilisation occur within the HLW classification, we shall focus on that area and concentrate on elements that are routinely classified as HLW, such as Cs, Sr, actinides and other longer-lived fission products.

Now that we have defined what we mean by HLW, what are the prevailing issues with the design and immobilisation of nuclear waste? To answer this question, we need to think back to what processes can occur with radioactive isotopes. If we look back to chapter 1, we can see that the following possibilities can arise:

(i) alpha-decay (α-decay)—the ejection of an alpha particle, a ^4He nucleus, from the nucleus of the decaying atom;
(ii) beta-decay (β-decay)—the ejection of an electron from the nucleus, arising from the internal conversion of a neutron to a proton;
(iii) beta$^+$ decay/positron emission—the inverse of beta-decay, it results from the conversion of a proton to a neutron, giving rise to the release of an anti-proton, or positron;
(iv) electron capture—a process similar to positron emission, but where an electron is captured from an inner orbital and converts a proton to a neutron;
(v) gamma emission (γ-ray)—emission of a quantised electromagnetic wave from the nucleus of the atom, arising from internal energy transfer within the nucleus;
(vi) spontaneous fission—where the nucleus spontaneously splits into two or more particles.

Since we are focusing on how to safely immobilise such isotopes, these processes need to be included, and the effects for three in particular need to be planned for. So, when designing a new waste form, what are the key criteria for a material that can be used in the design?

We can answer this question by focusing on the key criteria outlined below:

(i) waste loading—the waste form must be capable of accommodating acceptable levels of radioactive isotopes;
(ii) tolerance of impurities—the waste must be able to accommodate the radioactive isotopes and associated impurities;
(iii) stability—the material chosen must be thermodynamically and kinetically stable, i.e. it should not fall apart when being used.

We have now defined what HLW contains, described the processes that can occur with radioactive decay and set criteria that we can use to develop matrices for nuclear waste immobilisation. What are the possibilities that can be used?

Before we talk about the options available for nuclear waste immobilisation, we should consider answering the following question, as its answer will help greatly in designing nuclear waste options. Put simply, where are radioactive elements found naturally? The simple answer to this question is that they are found as minerals, so we should begin our discussion there.

7.2 Natural sources of uranium/thorium

Most uranium/thorium is found in the minerals uraninite (UO_2), pitchblende (a mix of UO_2/UO_3) and coffinite $(U,Th)[(OH)_{4x}|(SiO_4)_{1-x}]$, whereas thorium can also be found within monazite $(Ln,Th)PO_4$ sands. If we think back to when we discussed the nature and behaviour of uranium as an oxide in fuel, we can already see some commonality; for example, UO_2, when used as a fuel, can oxidise to UO_2/UO_3. There are other minor sources for both U and Th, and we shall introduce some of them later in our discussion.

As we are using U and Th, there will be radioactive decay present, since they both undergo alpha-decay. This decay can be used to determine the age of minerals, since we know the radioactive half-lives of U and Th isotopes. Could we use such minerals to develop a waste form?

The answer to this is yes, and for two reasons, with the first being obvious—the minerals are present, so we can examine how they behave. The second reason is less obvious, and concerns the first terrestrial nuclear reactor.

7.2.1 Oklo—a natural nuclear reactor

While it is often assumed that the first nuclear fission reactor on the planet was at Stagg Field (1942, see chapter 4), this is not true. The first reactors were in Africa ~1.5–1.7 billion years ago, in the Oklo region of the Gabon, which is now used as a uranium mine. The process by which the reactors were discovered highlights the nature of the nuclear industry and how it can lead to interesting results.

In 1972, an analysis of mined material was undertaken at the Pierrelatte nuclear facility [2], and it found that that the material was depleted in ^{235}U. Natural uranium contains 0.7202% ^{235}U and the samples were lower than this, at 0.7101%. On checking, it was found that there were other samples going back to 1970 that were even lower, at 0.44%, giving a total shortage in expected ^{235}U of 200 kg in 700 tons. This was a significant amount of material, needing an explanation and further examination.

Analysis of the isotopic composition [2, 3] for the elements present in the ore came up with other discrepancies, specifically in the isotopic composition of Nd and Ru:

(i) natural Nd contains 27% ^{142}Nd, and at Oklo this was <6%, with it being higher in ^{143}Nd;

(ii) natural Ru contains ~12.7% ^{99}Ru, and at Oklo this was 27–30%.

The conclusion drawn from these results was that nuclear fission had occurred in the past, specifically 1.5–1.7 billion years ago. Such a conclusion gives rise to two questions: how did it happen then, and can it happen now?

Before we answer the first question, we can easily answer the second one by looking at the half-lives and current abundance of ^{238}U and ^{235}U, as shown in table 7.1, and calculating the amount of ^{235}U present 1.6 billion years ago.

Table 7.1. Natural abundances for ^{238}U, ^{235}U and ^{239}Pu.

Isotope	% Abundance	Half-life
^{238}U	99.284	4.468×10^9
^{235}U	0.72	703.8×10^6
^{239}Pu	0.000	24.1×10^3

In chapter 1, we discussed radioactive decay, and in particular the equation that describes how decay behaves, shown here. If we rearrange the equation, we can obtain the amount of an isotope present in the past:

$$A(t) = A_o e^{-\lambda t} \Rightarrow A_o = \frac{A(t)}{e^{-\lambda t}}. \tag{7.1}$$

Now, using the information in table 7.1 to calculate λ,

$$\begin{aligned} \lambda t &= \left[\frac{\ln 2}{703.8 \times 10^6}\right] \times 1.6 \times 10^9 \\ \lambda t &= 1.58 \end{aligned}. \tag{7.2}$$

If this value for λ is then entered into (7.1), we obtain the following:

$$A_o = \frac{A(t)}{e^{-\lambda t}} = \frac{0.72}{e^{-1.58}} \approx 3.5\%. \tag{7.3}$$

Thus, 1.6 billion years ago the ratio of ^{235}U:^{238}U was significantly different, and if we compare the amount of ^{235}U present then with the amount present in reactor fuel used today, they are comparable. This differential half-life also explains why there is no ^{239}Pu found naturally on the planet. So, while the ratio of ^{235}U:^{238}U was much higher in the past than now, does that mean that a naturally occurring nuclear reactor definitely occurred?

The answer to this question requires us to consider the key criteria for sustainable nuclear fission:

(i) a source of fissile material;
(ii) low neutron absorption within core;
(iii) ideally, moderation to enhance neutron induced fission;
(iv) optimal geometry for maximum fission efficiency.

Using (i) and (iii) as key criteria, in modern reactors we use enriched nuclear fuel, i.e. ~5% ^{235}U as the fuel, and water as both the moderator and coolant. Therefore it is possible, and indeed was envisaged by Kuroda in 1956 [4], when he predicted the potential for a naturally occurring reactor. One thing, however, was overlooked, the moderating abilities of H_2O, which is used within modern reactors to increase the efficiency of neutron-induced fission.

So now that we know what is required for a natural reactor, what happened at Oklo? To paraphrase Goldilocks and the three bears, the conditions 'were just right' about 1.5–1.7 billion years ago, from which period 16 reactors have been found, running for ~2.5 h and using 6 t of ^{235}U. So how did they operate?

The geology at Oklo provides a mechanism for allowing the reactors to run, with layers of sandstone surrounding the ore, above a base of granite. These layers of permeable sandstone allowed water to permeate into the ore, slowing the neutrons from spontaneous natural fission, increasing the likelihood of capture, and inducing further fission. While the reactors operated, and the temperature rose to ~573 K, this water would rise out of the ore, thus reducing the moderation and slowing the

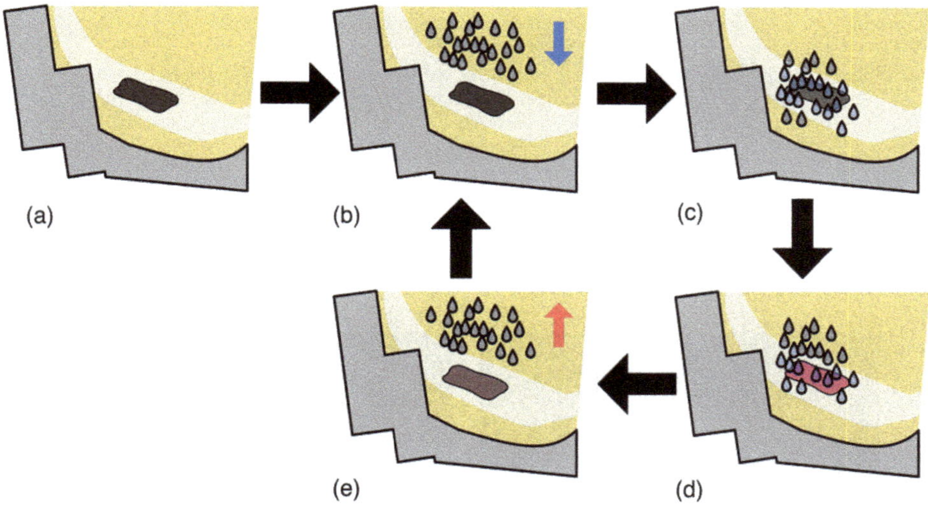

Figure 7.1. Schematic of the process and geology at the Oklo natural reactors: (a) mass of ore at the beginning, (b) water descending down to the core, (c) initiation of fission, (d) fission within the ore and (e) water heated and moved away from core, slowing fission down until the water cools.

reactor. As the water cooled in the sandstone above the ore, it would permeate back down in to the ore and repeat the process (figure 7.1).

Whilst the formation of a naturally occurring reactor is fascinating, it is the outcome and legacy that concerns us here. Since the reactor underwent fission, similar fission/decay products to those formed in modern reactors would result. At Oklo, ~5.4 t of fission products and ~15 t of Plutonium and other transuranics formed. How these behaved provides valuable insights into materials for the waste produced by modern reactors. In a report by the US government during the development of the Yucca mountain repository, the following statement was made [5]:

'And when these deep underground natural nuclear chain reactions were over, nature showed that it could effectively contain the radioactive wastes created by the reactions. No nuclear chain reactions will ever happen in a repository for high-level nuclear wastes. But if a repository were to be built at Yucca Mountain, scientists would count on the geology of the area to contain radionuclides generated by these wastes with similar effectiveness.'

Using the lessons learnt from Oklo and the natural occurrence of U/Th in minerals, we shall now focus on the development of ceramic hosts for high-level waste and the associated challenges.

7.2.2 Development of ceramic waste forms

Rather than developing a new material from first principles, if we were to base a matrix for the immobilisation of U/Pu/Act on a mineral that already contains U/Th, we could more readily develop the ideal system. Table 7.2 shows minerals that have been found to contain U/Th and which could be used as a basis for development, as

Table 7.2. Natural mineral sources known to contain U/Th.

Mineral	Composition
Zircon	$ZrSiO_4$
Pyrochlore	$(Na,Ca)_2Nb_2O_6(OH,F)$
Monazite	$LnPO_4$
Zirconolite	$CaZrTi_2O_7$
Fluorites	CaF_2 (UO_2, PuO_2, ThO_2, ZrO_2)

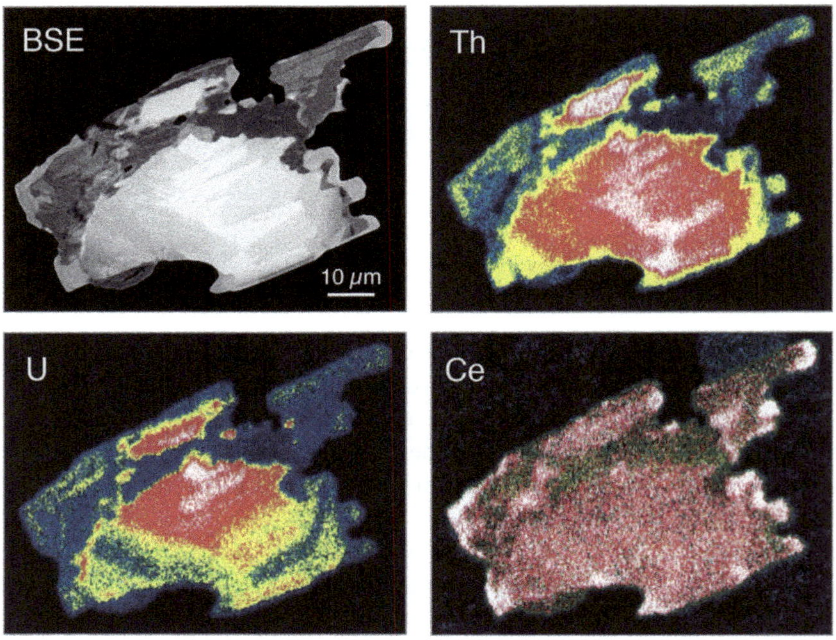

Figure 7.2. Electron micrograph showing U, Th and Ce distribution in a zoned zirconolite [6].

Table 7.3. Natural mineral sources known to contain Cs/Sr.

Mineral	Composition
Pollucite	$CsAlSi_2O_6$
Celestite	$SrSO_4$
Strontiate	$SrCO_3$

shown in graphically in figure 7.2. We can also use the same basis for Cs and Sr and start to obtain a range of potential hosts (table 7.3).

We can now use these minerals, together with the criteria we outlined previously—a high level of loading, tolerance for impurities and high stability—to develop potential waste forms. One of the earliest ceramics proposed was supercalcine [7, 8].

7.2.3 Supercalcine

This proposed waste form, based on silicates, molybdates, aluminates and phosphates, was founded on the premise that after reprocessing the oxides of Si, Ca, Al and Sr would be added and heated at 1473 K for ~2 h. This heating would give rise to a range of phases, as shown in table 7.4.

The main advantage of supercalcine is high waste loading (~70% wt), but its main disadvantage is that its solubility in water is similar to that for borosilicate glasses. Since we have set a criterion of high stability for use, supercalcine fails to meet our goal. What can we do to increase our chances of developing a better waste form?

The best way to answer this question is to apply what we know about the chemistry of the actinides and compare them to existing systems based on the lanthanides, which can have similar chemical behaviour.

7.2.4 Designing ceramic waste forms

If we compare the ionic radii of actinide and lanthanide elements, an interesting feature becomes apparent that can be used to develop waste forms. The radii for actinides with an oxidation state of 4+ are similar to that for a lanthanide with an oxidation state of 3+ [9]. Therefore, we can mix and match elemental composition to optimise waste loading, using one of the methods below:

$$\begin{aligned}
\mathrm{Act}^{4+}+\mathrm{M}^{3+} &= \mathrm{Ln}^{3+}+\mathrm{M}^{4+}\\
\mathrm{Act}^{4+}+\mathrm{M}^{4+} &= \mathrm{Ln}^{3+}+\mathrm{M}^{5+}\\
\mathrm{Act}^{4+}+\mathrm{M}^{2+} &= \mathrm{Ca}^{2+}+\mathrm{M}^{4+}\\
\mathrm{Act}^{4+}+\mathrm{M}^{3+} &= \mathrm{Ca}^{2+}+\mathrm{M}^{5+}\\
\mathrm{Act}^{4+}+\mathrm{Ca}^{2+} &= 2\mathrm{Ln}^{3+}\\
\mathrm{Act}^{4+} &= \mathrm{Zr}^{4+}.
\end{aligned}$$

To improve the nature of the material, we can add additives, and since we wish to avoid an Oklo-type event occurring, neutron poisons such as Gd, Hf, Cd and B. Using this methodology led to the development of one of the most important families in ceramic waste immobilisation, and the development and optimisation of synthetic rock, better known as Synroc.

Table 7.4. Mineral composition of supercalcine.

Mineral	Base composition
Apatite	$Ca_5(PO_4)_3OH$
Pollucite	$CsAlSi_2O_6$
Monazite	$LnPO_4$
Scheelite	$CaWO_4$
Powellite	$CaMoO_4$
Spinel	$MgAl_2O_4$
Fluorite	CaF_2

7.2.5 Synroc

The premise behind Synroc [10, 11] is a titanate-based ceramic capable of immobilising principally actinides, caesium and strontium. The initial compositions within Synroc were based on a range of mineral hosts, but over time this has settled into a multicomponent mixture based on fluorites (zirconolite/pyrochlore), perovskites and hollandites, with TiO_2 also being present. Since both zirconolite and pyrochlore are structural modifications of fluorite, and they can both be found within Synroc, for the purposes of this discussion we shall use the fluorite designation. The designed elemental distributions are shown in table 7.5.

As with supercalcine, Synroc was designed to immobilise multiple elements at the same time, with the elements designed to locate within the desired system, e.g. Cs in hollandite. Synroc showed that multicomponent waste forms can be designed and implemented, and has formed the basis of other similar systems. An example image of a Synroc ceramic is shown in figure 7.3.

There is one topic we have not yet considered, what would happen if we had enough fissile material to induce fission?

Table 7.5. Designed elemental composition within Synroc. The oxidation state for each element is indicated. Act refers to actinide elements and Ln to lanthanides.

Fluorites	Perovskites	Hollandites
U^{4+}, Th^{4+}, Pu^{4+}, Cm^{4+}, Am^{4+} and Np^{4+}	Pu^{3+}, Am^{3+} and Cm^{3+}	Cs^{+}, Rb^{+} and K^{+}
Act^{4+}	Act^{3+}	Ba^{2+}, Fe^{2+} and Ni^{2+}
Ln^{3+}	Ln^{3+}	Cr^{3+}
Sr^{2+}	Sr^{2+}, Na^{+}	Mo^{4+}

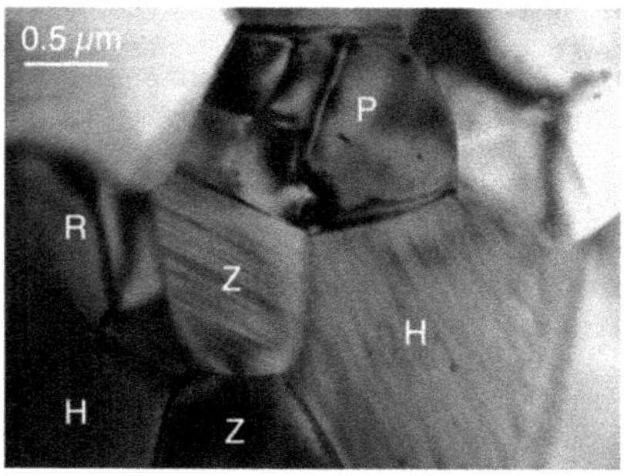

Figure 7.3. Transmission electron microscope image showing the microstructure of a multiphase ceramic and the phases present. Z = zirconolite, H = hollandite, P = pyrochlore and R = rutile. Reproduced with permission from [12].

7.2.6 Criticality prevention

The potential for fission to occur in a waste form in a similar way to that at Oklo has been considered and measures have been developed to prevent such an occurrence in a repository. To prevent such criticality, most waste forms now include some neutron absorbers, such as Gd, B, Hf and Cd. For example, the waste form developed for the Yucca mountain facility by Lawrence Livermore National Laboratory [13] was based around using pyrochlore as the host, but with Hf and Gd added, i.e. $Ca_{0.89}Gd_{0.22}Hf_{0.23}U_{0.44}Pu_{0.22}Ti_2O_7$, with excess TiO_2 and HfO_2 added.

7.3 Long-term effects in waste forms

In many ways, the challenges that impact on nuclear waste materials are similar, if not identical, to those affecting materials used within a reactor core and as fuel, such as radiation damage and transmutation. For the purposes of this discussion, we will use a general ceramic for discussion and link with examples where the effects have been examined.

7.3.1 Radiation damage

In a nuclear waste form, a significant proportion of the active components will be undergoing alpha and beta-decay, with a small amount of spontaneous fission, and with each process inducing different levels of damage. By far the greatest source of damage, however, arises from alpha-decay and, as such, we will focus on this.

Alpha-decay, as outlined in chapter 1 and earlier in this chapter, is the ejection of an alpha particle (^4He nucleus) from the nucleus, giving rise to two particles, each having momentum and being capable of inducing damage (figure 7.4). In a manner similar to a gun firing a bullet, the light atom (bullet) moves rapidly, while the heavy atom (gun) recoils. In chapter 2 we examined how radiation damage is induced, and discussed how the energy of a particle influences the damage it causes. In an alpha-decay event the two particles can have a range of energies, but if we use ~4 MeV for the He nuclei and ~90 keV for the recoiling atom, we can begin to predict the effects from each particle.

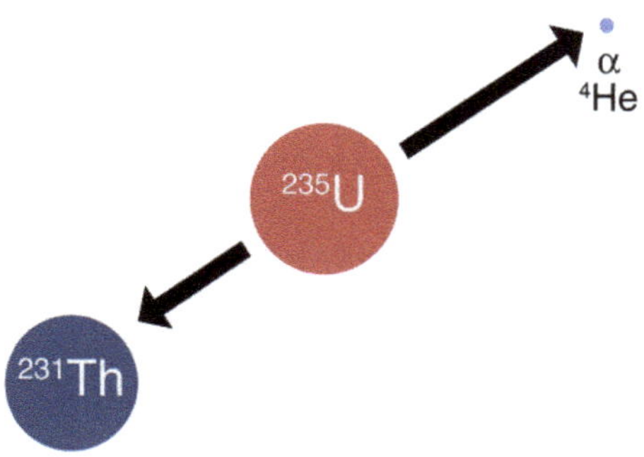

Figure 7.4. Schematic of the two particles arising from alpha-decay.

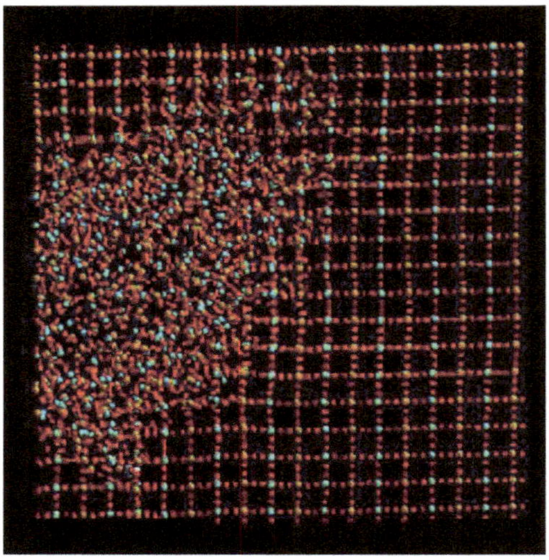

Figure 7.5. Simulation of two overlapping radiation damage events, leading to a larger volume of damage [14], see also animation.

Alpha (He-nucleus) particle

Since this particle is light and moving rapidly through, using the Kinchin–Pease methodology we developed in chapter 2, it primarily transfers energy through electronic excitations. As the energy of the particle decreases, it increasingly interacts with the lattice through collisions. This process gives rise to relatively low volumes of damage over an extended range within the lattice, which can often be recovered. However, the recoiling heavier particle behaves differently.

Recoiling particle

As this atom is much heavier, and much lower in energy, applying the Kinchin–Pease methodology, it will interact primarily through nuclear interactions, i.e. collisions, with significantly more damage—an example is shown in figure 7.5. While it is over a smaller range, the volume of damage is significantly higher than from the He-particle, and the lattice often does not recover.

In many ways, nuclear fuel and waste forms experience the same challenges, and we can consider them together. However, two significant differences are the rate of induced damage and the temperature of the waste form. The rate of induced damage is much slower in a waste form than within a nuclear fuel, and the temperature is significantly lower than that of fuel in operation. As such, the build-up of, and recovery from, damage can occur at a much lower rate.

So, how do these two processes impact on the waste form in the long term? The answer to this is complex, but we have already discussed what the effects could be in previous chapters.

7.4 Long-term behaviour of nuclear waste

One advantage of using ceramic-based waste forms is that in many cases there are natural analogues available, which can used to examine the long-term behaviour of radiation damage. Earlier, we spoke about minerals such as zircon, pryochlore and zirconolite being known to contain U/Th, and so we can use these to understand the long-term effects of damage and storage better.

7.4.1 Bulk effect of amorphisation

In many cases, natural mineral analogues are found to be amorphous, or metamict, for example, Rios *et al* [15] showed that as the mineral zircon increases in age, the degree of damage increases, until the material becomes amorphous (figure 7.6). In chapter 2, we also showed that this form of damage build-up followed the direct impact method.

Zircon is the not only material to undergo such a process, pyrochlores and zirconolites also experience a similar process. Generally, amorphisation induces a volume expansion in the material, in the case of zircon such an expansion can be up to 18%, a rather significant expansion. In others, such as pyrochlore the expansion is much smaller (~5%). However, even small expansions can be problematic.

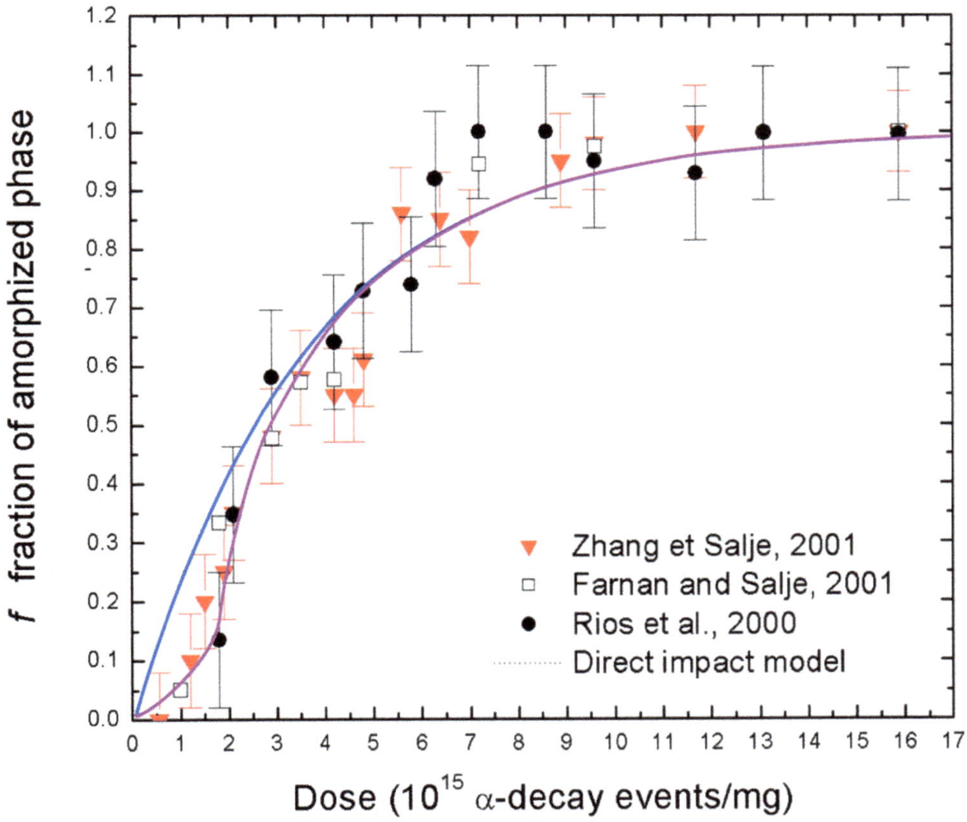

Figure 7.6. Fraction of amorphous phase within zircon as a function of damage, in this case defined as alpha-decays per mg of sample, with a direct impact model overlaid (purple line) [15].

Consider a multi-component ceramic waste form such as Synroc, with phases in close proximity to others, e.g. pyrochlore and hollandite. An expansion in one phase could induce cracks within itself or in adjacent phases. Hence, in order to minimise such problems, where possible, materials with a large volume expansion are replaced by those with a much lower expansion.

As with the effects we discussed in chapter 2, waste forms can still be expected to experience the same effects as other radiation-damaged materials. The same basic principles apply, and the Kinchin–Pease methodology for predicting types of damage can be used to help predict the effects.

In glass-based waste forms, the effects of damage are different, for example, one would not expect an amorphous material (glass) to become more amorphous. Such an idea becomes a philosophical discussion, and would also be difficult to quantify. However, similar to metallic alloys, in glass there is the possibility of radiation-induced segregation. This process can be accommodated through a modification in composition and is now a problem that can be handled.

So far, we have discussed what the effects of alpha-decay damage would be, and how they relate to those found in other nuclear materials. There are, however, two effects in waste forms that can be more pronounced and cause significant problems in waste forms. These are helium bubble formation and the effects of transmutation from beta-decay. Before we discuss transmutation, we shall investigate why gas bubble formation is problematic in waste forms and discuss the issues the bubbles introduce.

7.4.2 Helium bubble formation

Since alpha-decay, by its very nature, gives rise to a $^{4}He^{2+}$ nucleus, which picks up two electrons to become He, there is the potential, similar to that found in nuclear fuel, for He gas bubble formation. Historically, while helium was first discovered in the Sun, terrestrially it was found within uranium-containing ores. This helium can give rise to multiple problems, in many cases identical to those it induces within fuel, i.e. it is a source of stress/strain and locations where cracks can initiate. Why would cracks be a problem? To answer this question, we need to consider a likely storage scenario for nuclear waste.

7.5 Geological disposal of nuclear waste

In many countries, the proposed method for the long-term disposal of nuclear waste is underground. There are multiple options available to achieve this, but for the purposes of this discussion we shall use a generic facility underground, e.g. a geological disposal facility (GDF).

When considered more generally, the geological disposal of nuclear waste uses a multi-barrier concept, where the waste form chosen is one part, and is surrounded by a container, often within a clay barrier (bentonite), with this finally being placed within a concrete lined hole. Such a concept provides for multiple redundancy in the safe storage and prevents the release of material. However, even though a release is unlikely, as we established when discussing reactor design, we still need to address the potential for problems. One such problem is leakage of groundwater into the facility.

7.5.1 Corrosion and nuclear waste

It is very rare for a suitable location for a GDF to be completely dry, and since the likely storage time for some nuclear waste is significant, for example, the Yucca mountain facility has a design life of 100 000 years, any ingress of water can be a problem. Therefore, where possible, barriers are placed to prevent water ingress into the waste, hence the multi-barrier concept. However, to ensure safety we still need to consider any possible effects, and mitigate or minimise them.

In many ways, the dissolution of nuclear waste follows similar mechanisms to those for nuclear fuel/cladding, as we discussed in chapter 5. The exact chemistry is different, but we would still expect some dissolution to occur, with some systems being more soluble than others. In general, a stable oxide will form on the surface of the waste form; for example, in Synroc, rutile, anatase, zirconolite and hollandite will often form (see figure 7.7 for examples) [16]. However, the formation of a surface layer is not the concern, what is of more concern is the dissolution of the host matrix and the subsequent transportation of radioactive isotopes.

Now consider a generic material undergoing a dissolution process, it generally occurs from the outside in, and if we introduce a crack into the material, the dissolution can now occur from within it, increasing not only the rate of dissolution, but also potentially any leaching of radioactive isotopes. Since one key factor in determining the rate of dissolution is surface area, the more surface area there is, the more rapid the dissolution is, i.e. it follows standard reaction kinetics.

So, if we have He gas bubbles forming in the material, not only are we introducing an increased potential surface area for leaching, but, when this is combined with the increase in local stress, we are also increasing the rate of stress corrosion cracking. This process works in a similar manner to the stress corrosion cracking (SCC) experienced by both nuclear fuel and cladding. If we consider that cladding forms oxides on the surface when it undergoes SCC, which can lead to issues with hydride formation, in waste forms the counterpart could be the preferential leaching out of radioactive isotopes. An example of such a depletion can be seen in figure 7.8 [17].

If we have the potential for such issues when disposing of nuclear waste, is there anything that can be done to minimise the effects? Essentially the answer is yes, one

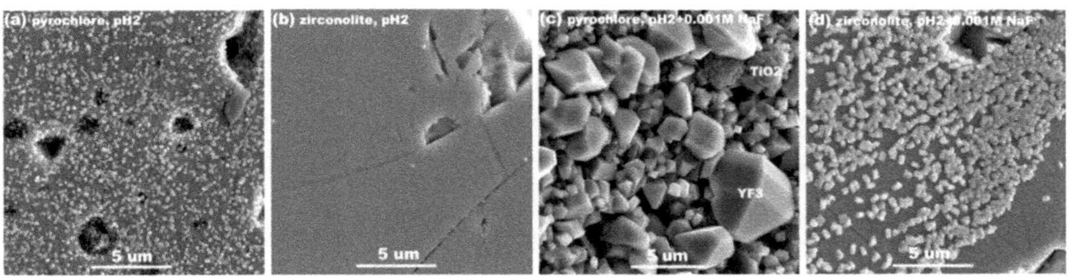

Figure 7.7. SEM micrographs after dissolution testing at 150° C for four weeks: (a) pyrochlore in pH2, (b) zirconolite in pH2, (c) pyrochlore in pH2 + 0.001 M NaF and (d) zirconolite in pH2 + 0.001 M NaF [16].

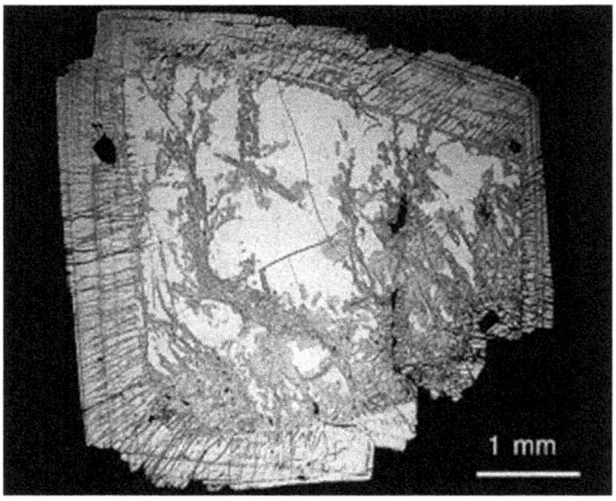

Figure 7.8. Example of a U depletion zone in a zircon from Ontario, showing the links between damage, expansion and corrosion. Reproduced from [17].

of the major issues in deciding the location of repositories is the local geology being compatible, i.e. it must be able to support the building of the facility. However, once we know the local geology, water conditions and likely ingress, we can design the ideal waste hosts. As with the development of nuclear fuel and cladding, what is ideal may not be practicable, but design changes can minimise any potential problems, and can often remove them entirely.

So, if we have issues with ceramics dissolving in water, how do they compare with glasses—in fact, how do ceramics and glasses compare more generally?

7.6 Ceramics and glasses—comparison

If we are comparing glass- and ceramic-based waste forms, what would be good metrics to use, are there any that would win out over others? For example, we could use as a metric how much active waste we could place inside the matrix, or how each material dissolves when tested.

Taking each system and listing its advantages and disadvantages, we can come up with a list that is not a complete one, but does show some of the advantages/disadvantages each has.

7.6.1 Ceramics

The advantages from using ceramics as hosts are:

(i) high loading of waste—it can be very high in some cases;
(ii) predictable material properties—initially at least, which are ideal for the application;
(iii) chemical and thermal stability—generally they are the most thermodynamically stable with an extended range of stability at temperature;
(iv) ability to use mineral analogues for examination of long-term stability.

The disadvantages are:

(i) increased cost in manufacture—new technology, such as hot isostatic pressing (HIP), is used;
(ii) restrictions in chemical flexibility—must be able to form the ceramic;
(iii) the effects of damage—volume expansion can prevent ceramics being used.

7.6.2 Glasses

The advantages glasses have are:

(i) variation in composition—glasses can vary significantly in composition to match the desired application;
(ii) resistance to amorphisation—since they are amorphous, volume expansion is less of a concern;
(iii) cost of manufacture—since we have been making glass for centuries, the technology is mature and comparatively cheap.

The disadvantages are:

(i) lower waste loading—the glass must be able to flow during manufacture, and must also be able to contain the waste;
(ii) thermal stability—at temperature, glass will devitrify (crystallise), which can significantly modify its properties;
(iii) lack of long-term analogues—while the oldest glasses can be thousands of years old, minerals can be billions of years old;
(iv) chemical stability—glasses are generally not the most thermodynamically stable form.

Rather than using just one metric, it is better to use a range and compare them in totality, i.e. use all the metrics to decide which is best (table 7.6).

If we use the metrics in table 7.6, we can begin to decide upon a likely winner, or, as we will show, there really is not a clear winner, and it comes down to a matter of perspective.

Table 7.6. Metrics that can be used to determine which waste form type is best for an application.

Metric	Which is better?
Chemical stability	Ceramic
Mechanical properties	Ceramic
Waste loading	Ceramic
Processing/fabrication cost	Glass
Thermal stability	Ceramic
Flexibility	Glass
Resistance to radiation damage	Glass
Long-term behaviour analogues	Ceramic

Since we are comparing ceramics and glasses in totality, we can also include other criteria, such as any economic savings that can be made if we favour one over the other. For example, while the cost of fabricating ceramics over glasses may be higher, the increased waste loading can compensate, which in turn could mean a smaller disposal facility. Alternatively, the cost savings made from using glasses might allow them to be used if the repository is an already existing cave with significant excess volume, and hence the cost of building the GDF is not increased.

For each technology proposed, there are mechanisms available that can enhance them, for example, when using tailored ceramics, the waste from reprocessing can be used. An example of this would be to remove Cs/Sr from HLW and immobilise it in a specific ceramic, e.g. hollandite, and place the actinide isotopes in a fluorite-based host. Therefore, in order to justify a specific materials technology, the complete process needs to be included in the discussion.

However, if ceramics and glasses both have advantages, is it possible to use them in a way that enhances the advantageous properties, while minimising the deleterious ones? Since we are discussing nuclear materials, this is an ideal topic, and one that we can answer.

7.6.3 Glass–ceramics

The premise behind combining glasses and ceramics is to make use of the advantageous properties of each to improve the waste form. For example, a ceramic could be designed to store actinide waste, within a glass barrier that can be used as a protection against leaching. An example is shown in figure 7.9.

The development of glass–ceramic waste forms is rapidly becoming a more active area of research worldwide and, in a manner similar to the development of advanced reactor designs and fuels, it is the next focus for waste form research.

So we have outlined what definitions of nuclear waste are used, the lessons that can be learnt from Oklo to design nuclear waste, and how the new materials are being designed and improved. However, we have not yet answered the question that

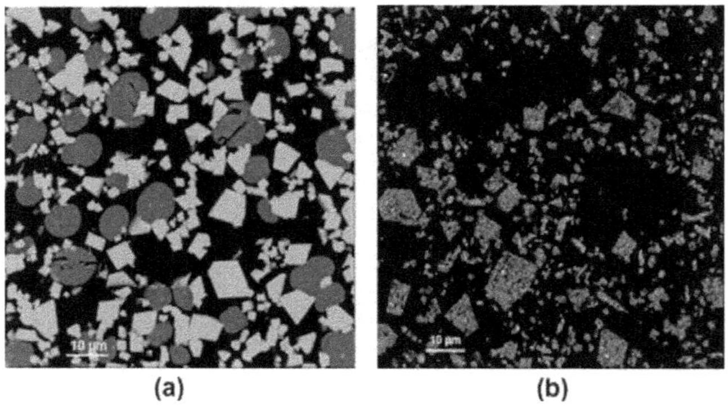

(a) (b)

Figure 7.9. Backscattered SEM images. (a) The U sample with glass (dark matrix), CaF_2 (round grey phase) and pyrochlore (bright crystals). (b) The U/Pu sample, containing glass (dark matrix), pyrochlore (grey phase) and trace amounts of $(U,Pu)O_2$ (bright spot) [18].

was proposed earlier in the chapter...the question of transmutation and how it effects waste form development.

7.7 Transmutation

The ability to convert base metals into gold was the aim of the alchemist, which led to the development of modern chemistry. Within the context of nuclear science, transmutation is easy, for example, we use transmutation to convert fertile isotopes into fissile isotopes, and use high neutron absorbers in fuel to prevent rapid burn in the early stage of operations. In these contexts, the process is advantageous, while in some others it is highly disadvantageous, and with regards to waste, it can be problematic.

When an element (e.g. ^{235}U) begins the journey down a decay series, the process is probabilistic in nature (see chapter 1), so with a large enough volume we can predict the isotopes/elements that will appear. For example, in the ^{235}U decay series we can predict that radon will form, and this is where one of the problems with transmutation arises, as a gas forms, which then decays back into radium, a metal. This could cause a problem if you were to breath in radon just before it decays into radium; your lungs would contain an atom of radium, which itself would undergo further decay. In some parts of the world, where granite is the local bedrock and contains uranium, e.g. Cornwall in the UK, radon gas build-up can be a significant problem.

So, we have had a look at one of the implications of transmutation, the change of state from solid to gas to solid, but a bigger problem can arise if when transmuting we also change the oxidation state of an element. This principally arises during beta-decay, electron capture and positron decay, but to make our life easier we shall refer to beta-decay in this discussion, while the principles can be applied to the other mechanisms.

7.7.1 Implications of beta-decay

As outlined in chapter 1, beta-decay is the emission of an electron from the nucleus, giving rise to a change in atomic number, which increases by one. This change can have significant implications, for example, the beta-decay of ^{137}Cs to ^{137}Ba (equation (7.4)), as any chemist will tell you, has also changed the stable oxidation state from +1 for Cs to +2 for Ba. This can have severe implications when developing materials to immobilise nuclear isotopes, and is only one of many that need to accommodated, for example, the decay of ^{90}Sr through ^{90}Zr involves a change in the stable oxidation state from 2+ to 4+ (equation (7.5)). So, how can we accommodate such a potential change in oxidation state?

$$^{137}_{55}Cs \rightarrow {}^{137}_{56}Ba + e^{-} + \gamma. \tag{7.4}$$

$$^{90}_{38}Sr^{2+} \rightarrow {}^{90}_{39}Y^{3+} + e^{-} \rightarrow {}^{90}_{40}Zr^{4+} + e.^{-} \tag{7.5}$$

One way is to design a waste form with elements that can vary their oxidation state in response, and such a system has been developed in the Synroc system, using the mineral hollandite.

Hollandite

Hollandite is a naturally occurring mineral with the general formulation $BaMn_8O_{16}$, with manganese having a variable oxidation state of 3+/4+ and a structure composed of channels (figure 7.10). The hollandite structure is capable of accommodating a wide range of compositions, and has a range of applications outside of nuclear power, but its main use is as the host of choice for Cs and Ba isotopes.

The hollandite structure, with a more generic composition of $A_xB_8O_{16}$, where A = Ba, Cs, and B = Ti, Al, Mn, Mg and others (figure 7.10), comprises tunnels made from edge- and corner-sharing B cations in octahedral co-ordination with oxygen. Within these tunnels, the Ba and Cs cations are located and ordered in a manner incommensurate with the unit cell (which is outside the scope of this discussion). The variation in composition on the B-site is designed to accommodate any change in charge state from the decay of Cs to Ba, with a reduction in oxidation of Ti or Mn from 4+ to 3+, as shown in equation (7.6):

$$Cs^{+} \rightarrow Ba^{2+}, Ti^{4+} \rightarrow Ti^{3+}. \tag{7.6}$$

This designed approach to chemical composition helps to mitigate any effects from transmutation, and should give rise to a more stable waste form. There is, however, another implication of transmutation that we can use beneficially, and which is indeed used in nuclear medicine.

Technetium

Technetium is one of only two elements below the atomic number of uranium that is not naturally occurring, with the other being promethium. They both arise from nuclear reactions, in the case of technetium as a fission product. Technetium can arise

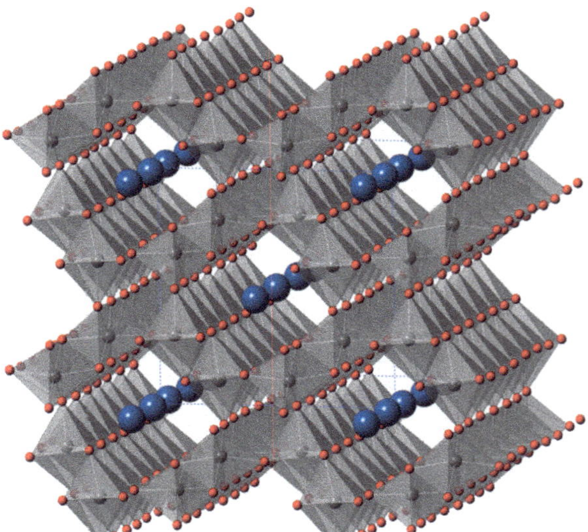

Figure 7.10. The hollandite crystal structure, showing the location of the Cs/Ba cations with respect to the tunnel [19].

in two ways, either directly as a fission fragment, or as a beta-decay product from molybdenum, with a range of isotopes. Technetium has multiple uses, but its main one is medical imaging and diagnosis, using ^{99m}Tc, which has a half-life of 6.25 h before decaying by gamma emission to ^{99}Tc, which has a half-life of ~250000 years shown in figure 7.11. So, if the half-life is only 6.25 h, how can we use it outside a reactor?

The answer to this question lies with transmutation and its mother isotope, ^{99}Mo. There are multiple methods by which ^{99m}Tc can be generated, but for regular use within hospital a generator based on ^{99}Mo is used, using the change in chemical state advantageously.

Since technetium and molybdenum are different elements, they can be expected to have different chemistries, and this is the key point in generating ^{99m}Tc for medical uses. For ease of discussion, we will not go through how the ^{99}Mo is generated—there are multiple methods available, from neutron bombardment through to extraction from nuclear fuel—and instead we will start the discussion at the extracted ^{99}Mo stage.

Using technology developed for the extraction of metals from a mixed solution, we can put ^{99}Mo on a column, for example, as a molybdate anion (MnO_4^{2-}) on a substrate such as alumina (Al_2O_3). This would be an ideal way of separating out Mo from any mixture, but in this case we are using it to fix ^{99}Mo, which undergoes beta-decay with a half-life of 66 h, forming ^{99m}Tc. Therefore, after 66 h, there is 50% MoO_4^{2-} on the column and 50% TcO_4^-, which does not stick to the column in the same manner and can be removed easily using a saline-based solution, forming soluble $NaTcO_4$, which can then be injected into the patient. A photograph of a Tc generator is shown in figure 7.12. The ^{99m}Tc can then be imaged from the gamma ray it emits before decaying down to ^{99}Tc, which is then ejected by the body.

There are two implications from using such a process. Firstly we are now using the change in chemical stability induced by transmutation advantageously to extract a usable isotope more efficiently. The other implication is less useful, since the half-life of ^{99}Mo is only 66 h, there is a need to replace the generators regularly, because after ~12 days the level of ^{99}Mo is low, and effectively unusable. If you couple this with the half-life of ^{99}Tc, which is 250000 years, hopefully you can see that we are slowly

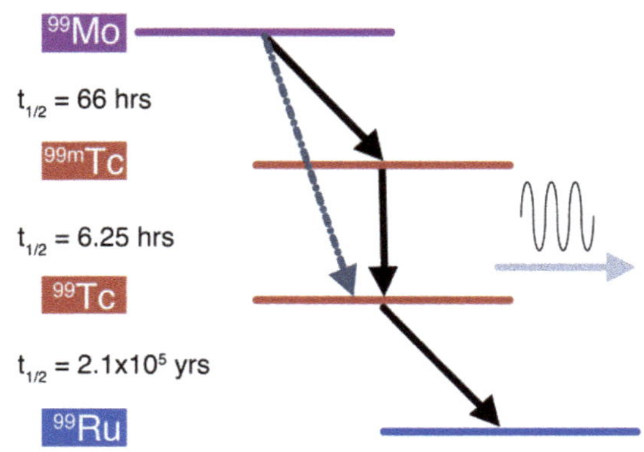

Figure 7.11. Decay series for ^{99}Mo to ^{99}Tc.

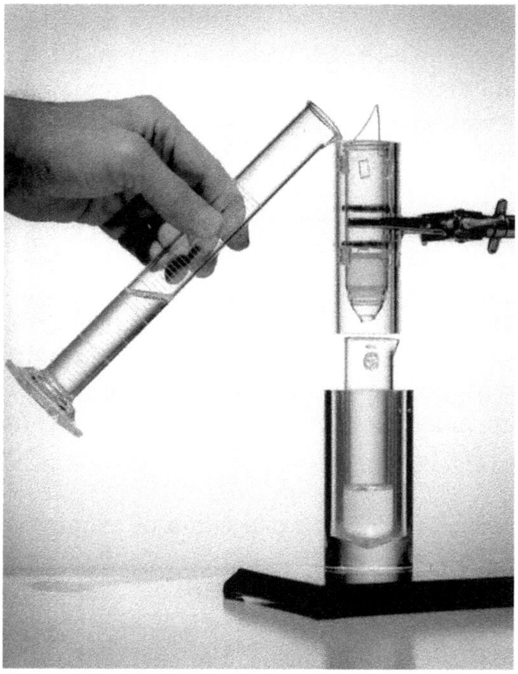

Figure 7.12. Image of a technetium generator, showing the elution of $NaTcO_4$ in solution [20].

building up a significant waste stream for ^{99}Tc and, given the increased use of such medical isotopes, this will increase further. This challenge will therefore need to be met by the development of waste forms that are ideal for technetium and it its decay products.

In this chapter, we have looked at how nuclear waste can be immobilized, and at the challenges that are faced. The rapid development of nuclear technology in the 1950s left significant challenges that needed to be overcome and now, with the modern methods being proposed, the safe disposal of nuclear waste can be achieved. In many cases, the same effects and criteria used for the development of materials/fuels for use within a reactor can be applied to the development of nuclear waste.

In the next chapter, we shall examine the challenges associated with the development of nuclear fusion, the alternative method for using the energy contained within the atom.

References

[1] IAEA *Classification of Radioactive Waste, General Safety Guide GSG*-1 (IAEA)

[2] Gowan G A 1976 *Scientific American* (July 1976) 36–47

[3] Gauthier-Lafaye F, Hollinger P and Blanc P L 1996 *Geochim. Cosmochim. Acta* **60** 4831–52

[4] Kuroda P K 1956 *J. Chem. Phys.* **25** 781–2

[5] US DOE *Oklo: Natural Reactors DOE/YMP-0010* (Office of Civilian Radioactive Waste Management, US DOE)

[6] Lumkpin G R, Smith K L and Gieré R 1997 *Micron* **29** 57–68

[7] McCarthy G J 1997 *Nucl. Technol.* **32** 92
[8] Roy R 1977 *J. Amer. Ceram. Soc.* **60** 350
[9] Shannon R D and Prewitt C T 1969 *Acta Crystallogr.* B **25** 925–46
[10] Ringwood A E, Kesson S E, Ware N G, Hibberson W and Major A 1979 *Nature* **278** 219
[11] Vance E R 2002 *J. Aust. Ceram. Soc.* **38** 48
[12] Lumpkin G R 2006 *Elements* **2** 365–72
[13] Plutonium Immobilisation Project UCERL-ID-133089 (PIP-99-012)
[14] Trachenko K O, Dove M T and Salje E K H 2001 *J. Phys.: Condens. Matter* **13** 1947
[15] Rios S and Salje E K H 1999 *J. Phys.: Condens. Matter* **11** 8947
[16] Zhang Z, Li H, Vance E R, Mcleod T and Scales N 2004 *Proc. Mater. Res. Soc.* **824** CC8.22.1
[17] Lumpkin G R 2001 *J. Nucl. Mater.* **289** 136–66
[18] Zhang Y, Zhang Z, Thorogood G J and Vance E R 2013 *J. Nucl. Mater.* **432** 545–7
[19] Whittle K R, Ashbrook S E, Lumpkin G R, Farnan I, Redfern S A T and Smith R I 2007 *J. Mater. Sci.* **42** 9379–91
[20] Brookhaven National Laboratory

IOP Publishing

Nuclear Materials Science

Karl Whittle

Chapter 8

Materials and nuclear fusion

So far, we have examined the principles and methodology of obtaining energy from the splitting of atoms, i.e. nuclear fission, and this chapter focuses on the alternative method for extracting energy through the fusing of atoms, i.e. nuclear fusion. There is currently one nuclear fusion reactor that we all use daily, the Sun, but the method it uses for fusion is difficult to replicate on Earth and, as we will discuss, there are significant challenges to overcome. These challenges are sometimes very similar to those faced in fission reactors, while others are at the extreme of what is currently achievable.

8.1 Atomic background and recap

In chapter 1, we discussed the nature of the atom and how energy can be extracted from the release of binding energy during fission. The principle behind fusion is similar and uses the same principles for estimating energy release. If we use the example in figure 8.1, which is shown schematically in figure 8.2, i.e. the fusion of deuterium and tritium, what would the expected energy release be?

To estimate the energy we can apply the same method we would use for a fission process, i.e. calculate the mass difference between the starting and final products and convert it to energy:

$$ {}^{2}_{1}\mathrm{H} + {}^{3}_{1}\mathrm{H} \rightarrow {}^{4}_{2}\mathrm{He} + {}^{1}_{0}\mathrm{n}. \tag{8.1} $$

As with nuclear fission, taking the atomic masses of ^{2}H, ^{3}H, ^{4}He and ^{1}n, we obtain the following:

mass of ^{2}H = 2.014102 amu
mass of ^{3}H = 3.016049 amu
mass of ^{4}He = 4.002602 amu
mass of ^{1}n = 1.00866 amu.

doi:10.1088/978-0-7503-1104-5ch8

Figure 8.1. Animation of the fusion process, in this case the fusion of deuterium and tritium to form helium and a neutron.

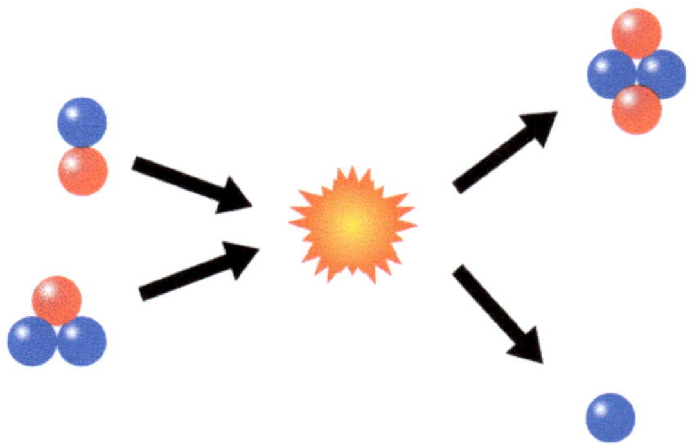

Figure 8.2. Schematic of the fusion process, in this case the fusion of deuterium and tritium to form helium and a neutron.

Therefore, the expected mass change would be:

$$\begin{aligned} \text{mass change} &= {}^{2}\text{H} + {}^{3}\text{H} - {}^{4}\text{He} - {}^{1}\text{n} \\ &= 2.014102 + 3.016049 - 4.002602 - 1.00866 \\ &= 0.018889 \text{ amu} \end{aligned}$$

$\therefore$ energy released = 0.018889 × 931.494 = 17.6 MeV.

As with nuclear fission, this energy release is across the two formed particles, in this case ^{1}n and ^{4}He, as 14.1 MeV for the ^{1}n and 3.5 for the ^{4}He. Later we shall discuss the implications of these particles/particles, but for now it is sufficient to say that we have released energy from the fusion process. Is this always the case?

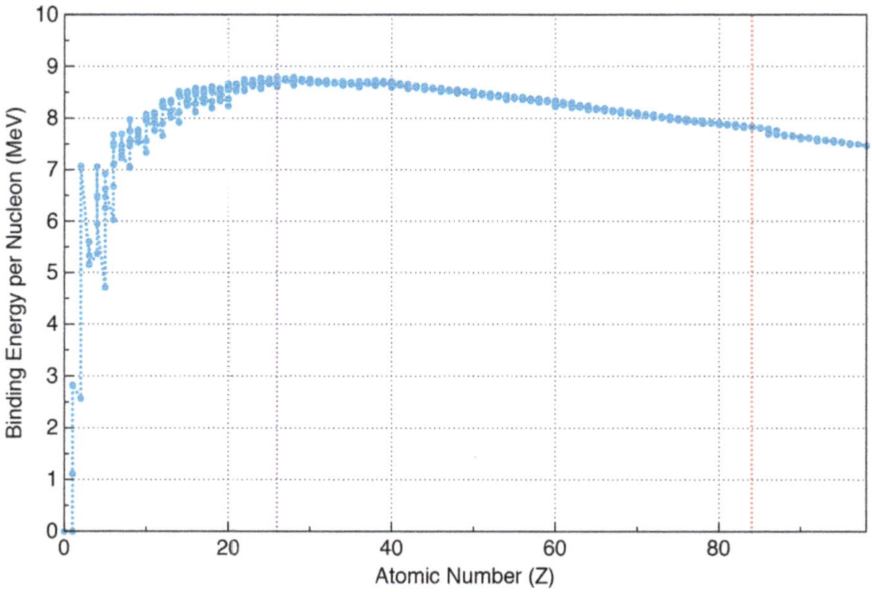

Figure 8.3. Diagram showing the relative stability per nucleon for the elements where data have been determined. Data taken from [1, 2].

The simple answer is no, as with the processes in the Sun, we can obtain energy from the fusing of atoms to make heavier ones. However, there comes a point when the process changes from being an exothermic to an endothermic process, i.e. more energy is put into fuse than is released by the fusion process.

We examined the concept of binding energy per nucleon (figure 8.3) back in chapter 1, and discussed how it related to fissionability, but here we need to examine it from an alternative viewpoint, the ability to form the atom.

As was discussed previously, the stability falls into three regions, and increases in energy until it reaches a maximum at Fe/Ni. After this, the energy begins to decrease, until at Po the probability of fission begins to increase. Since we are talking about fusion, the first region is of most concern to us.

The implication from the increase in binding energy per nucleon is that, in principle at least, it is possible to extract energy from the fusing of isotopes. Terrestrially, it is unlikely, for reasons we shall discuss shortly, that we will be able to fuse elements heavier than He, or possibly Li, given ideal conditions. However, while we are unlikely to fuse elements from Li through to Fe, it does happen in stars as they evolve, until they try to form Fe/Ni. At this point, the process moves from becoming exothermic to endothermic, requiring more energy to be put in than is released. A star achieves this energy input by collapsing and going supernova, something we may wish to prevent occurring terrestrially. So, if we can obtain energy from H and He fusing, why are we not doing it?

Put simply, fusion is not a trivial process, and to answer this question we need to consider the process of fusion and take note of how it occurs in the Sun.

8.2 Requirements for fusion

If we compare the requirements for fusion with those for fission, there are some commonalities. For example, there is the requirement for fuel, in the case of fission this is a fissile isotope, whereas for fusion it is an element which can be fused with another. However, the commonality then diverges, because while spontaneous fission can occur—we just increase the efficiency by inducing it with neutrons—in fusion the process is less simple and, put simply, spontaneous fusion does not occur. So, how do we fuse atoms, or, more specifically, nuclei?

The key to answering this question is to note the change in terminology used in the question, we are not fusing atoms, we are fusing nuclei, which are positively charged. Hence, we need to overcome the inherent Coulombic repulsion between two positively charged particles before we can even consider fusion. How could we overcome the repulsion?

If we take a star as an example, increasing gravity initiates the fusion process, which becomes self-perpetuating, with gravity providing a force for containment. During star formation gravity attracts particles together, principally hydrogen. As they are attracted together, they become more excited and increase in temperature. For comparison, a similar process happens when you compress gas quickly, the temperature and pressure increase as the volume decreases. As the atoms become more energetic, they are eventually able to form a plasma of charged species, i.e. loose electrons, which now have sufficient energy to overcome their mutual repulsion. At this point they can fuse, in this case giving rise to deuterium, tritium and helium. This is when the star begins and uses its available fuel (figure 8.4).

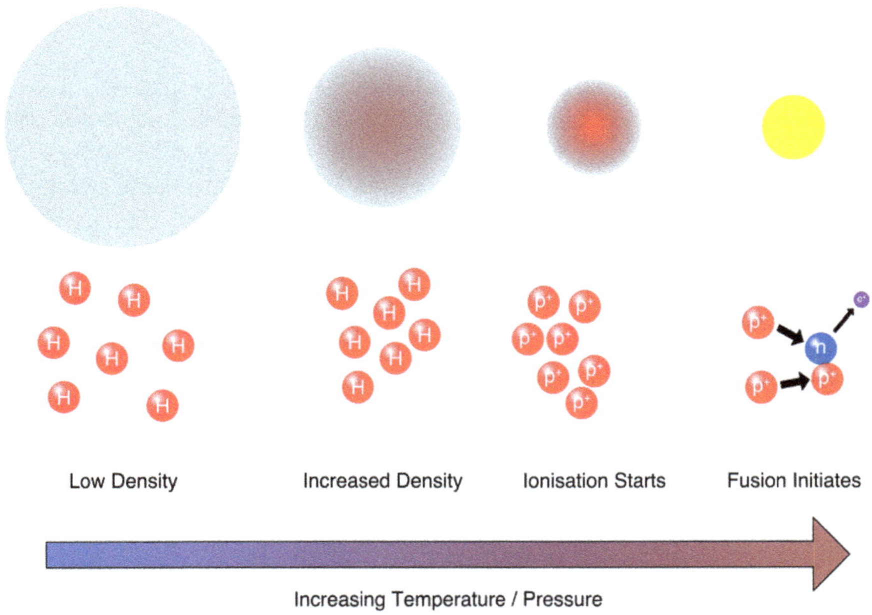

Figure 8.4. Schematic of fusion initiation in a star. It starts with HH fusion, which gives rise to deuterium and a positron (conversion of a proton to a neutron).

Terrestrially, we are currently unable to use gravity in a similar manner, but there are two alternative options available, magnetic and inertial confinement. While both these methods are being developed, much more research attention is focused on magnetic confinement. How does magnetic confinement work, and how does it compare to fusion within a star?

Rather than using gravity to contain the plasma, with magnetic confinement we use a strong magnetic field, which contains the plasma within a specific volume used for fusion.

Two designs are commonly studied to achieve this:

(i) tokamak—this uses a device that is similar to a doughnut in geometry, an example of this is JET at CCFE [3], and it is the basis of the ITER design [4];
(ii) stellerator—a Möbius strip designed for containment, examples of this are at Wendelstein (the 7-X [5]) and the University of Wisconsin-Madison (the Helically Symmetric eXperiment, HSX [6]).

Using ITER as the model for discussion, how would fusion be induced, and what are the material requirements?

8.3 ITER—the International Thermonuclear Experimental Reactor

ITER (Latin for 'the way') is a multi-partner facility, with funding from seven entities [4]. In alphabetical order, the funders are China, the EU, India, Japan, Russia, the US and South Korea. It is located near Cadarache in France. As outlined previously, the design uses a tokamak as its basis, and as such it has design requirements that outline the challenges involved in fusion material development.

8.3.1 Design of ITER

As can be seen in figure 8.5, ITER is a complex device, whilst in figure 8.6 a simplified model shows the key areas that can be used as a basis for the discussion of fusion materials.

Before discussing the material challenges, and how they are being overcome, we should first understand the outcomes from nuclear fusion, as they directly impact on materials within the core. These outcomes can then be coupled with the requirements for fusion, and as with nuclear fission, there are some incompatibilities that need to be recognized.

8.4 Outcomes and challenges in fusion

Earlier, we examined the fusion of ^{2}H and ^{3}H to form ^{4}He and ^{1}n (equation (8.1)), and considered the estimated energy released and the products formed, as shown in figure 8.7.

In figure 8.7 there are four boxes that highlight the areas that give rise to material challenges in both the design and operation of the reactor, providing an ideal example of the challenges in fusion materials development. However, we need to better understand the implications of each stage in the fusion process, and how they impact on the design of the reactor.

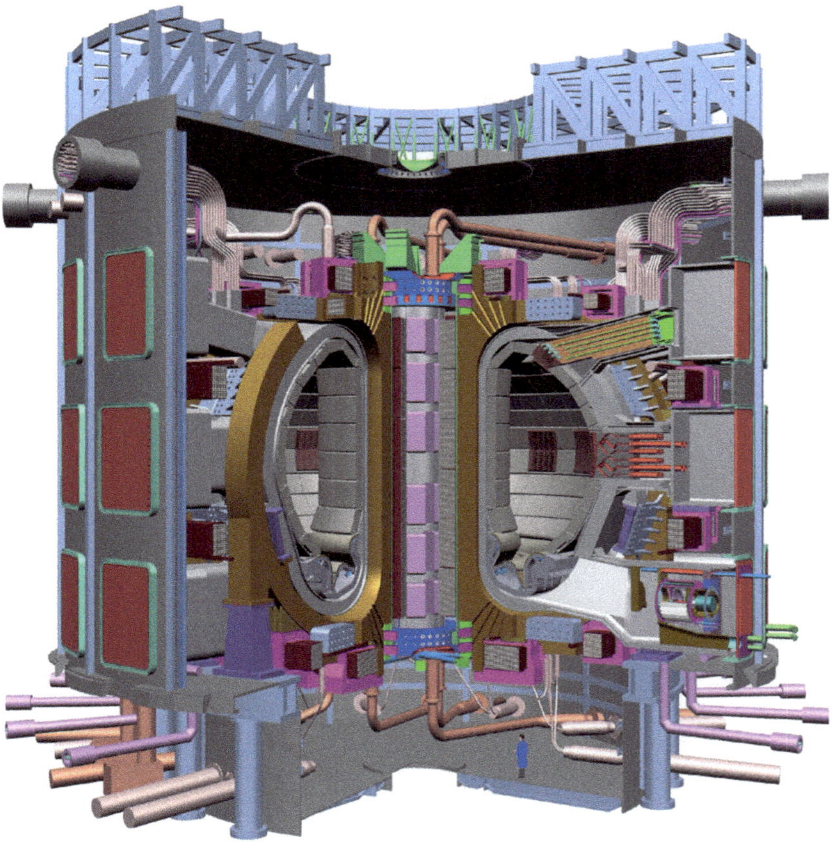

Figure 8.5. Schematic of ITER. A representative human is shown to indicate scale. Reproduced with permission from [4].

8.4.1 Fuel

As with fission, fuel is required for fusion to occur, and in the ITER roadmap the first experiments will fuse deuterium and tritium (D,T fusion) [4]. However, there is a problem that needs to be overcome. Tritium undergoes radioactive decay, with a half-life of 12.32 yr. As a consequence, very little is found naturally, with it mainly arising from nuclear fission using heavy water. In such a case, tritium is formed through neutron capture from deuterium, and this could provide a source. However, since the majority of water-based reactors use light water, the production of tritium is low, and whilst this is adequate for ITER, it will present a problem for the next reactor, DEMO. Therefore, a mechanism for generating tritium needs to be developed, as DEMO is expected to require ~300 g of tritium per day of operation [4], but luckily this is possible through the neutron-induced transmutation of other elements.

8.4.2 Heat

Since for fusion to occur we need high temperatures to overcome the atomic repulsion, and indeed heat is being released, the ability to contain the high

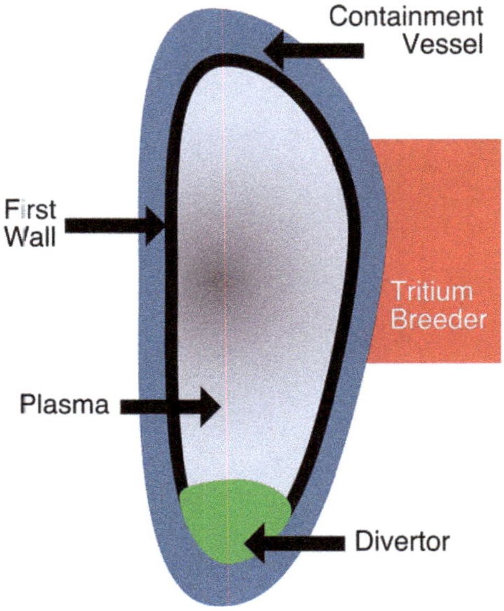

Figure 8.6. A simplified schematic of ITER, highlighting the regions of interest.

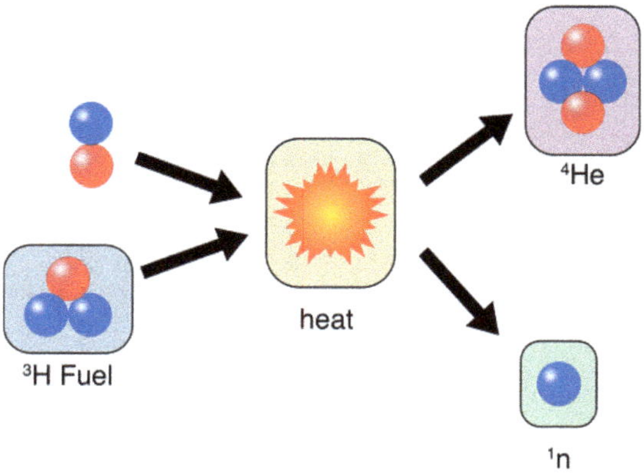

Figure 8.7. Schematic of the fusion of ^{2}H and ^{3}H to form ^{4}He and ^{1}n, with boxes highlighting the key challenges.

temperatures is vital—a temperature of ~150 000 000 K is expected within the plasma. Luckily, within the core the plasma pressure is low, outside the region where fusion is occurring, so heat is predominantly transferred by radiation, with little transferral by convection and conduction. However, even though the heat is being transferred by radiation, the first wall of the reactor is still expected to experience temperatures of 1000s of kelvin, and it will face >736 MW of heat [4]. This is not an easy challenge to overcome.

8.4.3 Helium

One of the products from D,T fusion is ^{4}He, with an energy of 3.5 MeV, which in this context is waste, as it cannot be fused easily and therefore needs to be removed.

Since the particles are charged, and ideally contained within the plasma using a magnetic field, it is possible to use the same magnetic field to drive them, i.e. divert them, to where they can be removed. This part of the vessel is the divertor (shown in figure 8.6), and it presents an extreme challenge in materials design.

8.4.4 Neutron production

The other product from D,T fusion is a neutron, with an energy of 14.1 MeV, and this will be used in two ways.

(i) As a means by which the energy released from fusion can be used to generate electricity. This will be primarily through the conversion of kinetic energy via a thermofluid, which will then be used to drive turbines.

(ii) In the generation of ^{3}H through neutron induced transmutation, i.e. the breeding of fuel.

Each of these uses requires the utilisation of neutrons, and so at least for the breeding of ^{3}H, it is a complex challenge.

8.5 Material requirements

We have examined the key stages of the fusion process, and discussed how they impact on the design, and we can now begin to draft a set of material requirements and use them to start selecting possible materials.

The key factors are as follows.

(i) Temperature—both that induced to allow fusion and that resulting from fusion. The induced temperature rise required to ignite the process uses magnetic induction.

(ii) Radiation damage—as has been described, both ^{4}He and ^{1}n are produced, each with momentum capable of inducing damage. The levels of damaged expected within the core are ~200 dpa, levels commensurate with some GenIV designs.

(iii) Chain reaction—the materials must allow the fusion process to continue, i.e. the chain reaction, similar to fission.

(iv) Tritium breeding ability—a mechanism that uses neutrons to induce transmutation is required, and therefore a material that allows neutron transport is needed.

Thus the key challenges are similar to those experienced in other contexts, e.g. nuclear fission and GenIV designs such as the VHTR/GFR. Could we use similar materials to the ones used for these?

In many applications we can often use the materials developed for high-temperature fission reactors as starting points, with the only caveat being that they must allow fusion to continue. Where are the key challenges for fusion materials, and how do these affect the materials?

If we use the schematic in figure 8.7 as a basis for discussion, the key areas are:

(i) first/plasma-facing wall;

(ii) magnetic containment;

(iii) the divertor;
(iv) the tritium breeder;
(v) tritium delivery.

8.5.1 First/plasma-facing wall

The key criterion for this first wall is the protection of other parts of the reactor. As a consequence, it must be compatible with some key criteria:

(i) heat—as outlined previously, it will experience (through radiation) very high temperatures;
(ii) magnetic confinement—we will discuss the materials used to provide the magnetic containment later, but the first wall must be compatible, i.e. not impinge upon the magnetic containment in a manner that is unpredictable;
(iii) plasma/fuel—the materials used here must have minimal impact on the plasma and fuel.

Combining these, we would end up with materials that have a high thermal conductivity and are non-magnetic. There are two main classes currently being investigated, and these are generally metallic or ceramic in nature.

The first metallic option, which is the method being used for ITER, is to use beryllium coated onto a copper-based alloy [4]. This method has advantages, for example, as both components are metallic, the thermal conduction should be high, and as they are both non-magnetic, it should not cause problems with plasma management. However, the main disadvantage would be the different thermal expansion co-efficients and crystal structures. If you remember back to when we were discussing nuclear fuel and cladding, a similar situation to that for an isotropic and an anisotropic material in close contact occurs here. While Cu can be expected to expand isotropically as it adopts a cubic structure, Be would be anisotropic, as it is hexagonal. As they are bonded to each other, when the temperature increases, an increase in interfacial stress/strain would be expected. Such an increase could lead to delamination, or failure of the material, so why has this been chosen? To answer this question, we need to understand the effects of the radiation more fully, and build upon what was discussed in chapter 2.

8.6 Radiation damage and the first wall

As discussed in chapter 2, there are many effects, with the inducement of stress/strain, the modification of material structure and amorphisation being just a few. There was one impact, however, that we did not discuss, and that is the possibility of sputtering from the surface.

When a particle impacts on the surface, there are three general types of interaction, and using the Kinchin–Pease methodology as a guide, the following would be expected to occur, with an example of damage shown in figure 8.8:

i) below E_d—this is where the incoming particle does not have enough energy to displace an atom;

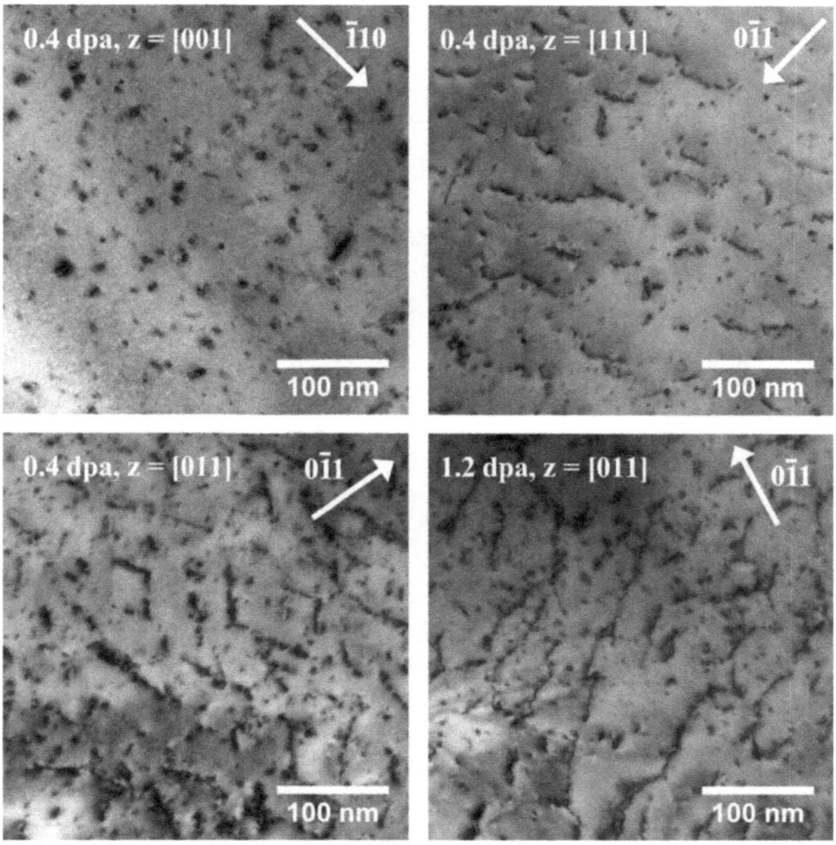

Figure 8.8. Electron micrograph of damage within W containing 5% Ta that has been irradiated with W ions at 300 °C to a damage level of ~1.2 dpa [7].

ii) above E_d—this is where the energy is sufficient to displace an atom and induce energy with nuclear interactions;
iii) above E_d—this is where the energy is high and the energy transfer is mainly through electronic interactions.

In general, the expected impacts from damage are predicted to come when the energy of the incident particle is above E_d, i.e. stress/strain, damage, amorphisation, etc. However, for the first wall within a fusion reactor, a problem that we have not considered yet—the removal of atoms from the surface through the process of sputtering—can arise, close to, or just below, E_d.

8.7 Sputtering

In many ways, sputtering can be thought of as the erosion of a material's surface, similar to sand blasting, where incoming particles knock particles off the surface, as shown in figure 8.9.

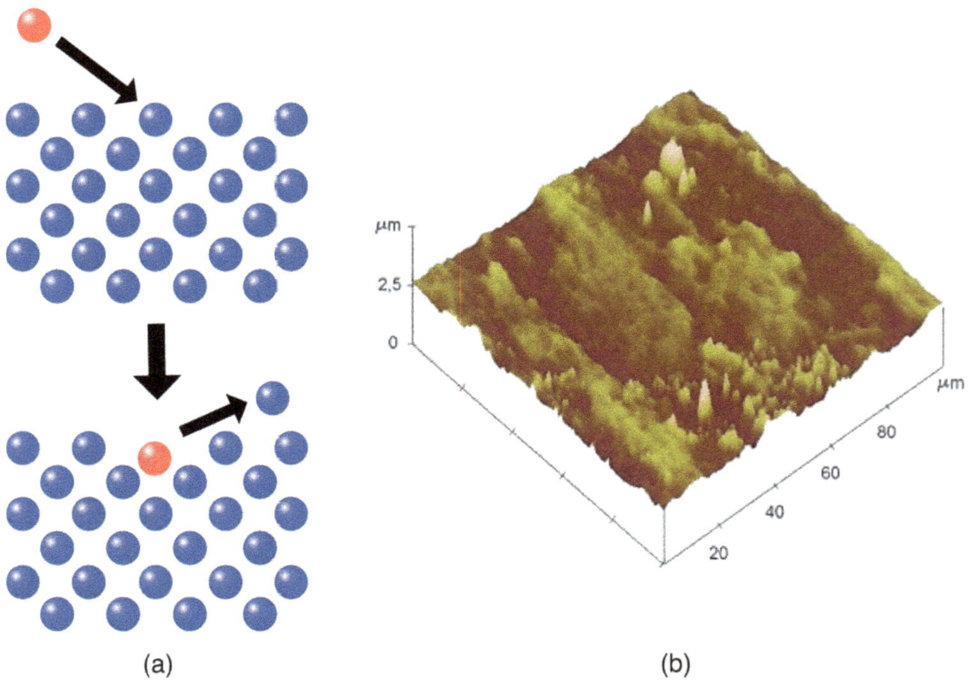

Figure 8.9. (a) Schematic of the sputtering of atoms from a surface. (b) AFM image of a tungsten ITER-grade specimen (3.0 dpa) after a layer of ~3.9 μm thickness has been sputtered with Ar ions from an ECR plasma. The imaged area is 100 × 100 μm^2 with a height scale span of 2.5 μm from [8].

In order for sputtering to occur, the incident particle must have sufficient energy to overcome the surface binding energy, which for metals is generally of the order of a few eV. The displacement energies for internal atoms are often much higher.

Sputtering can occur when the energy of an incident particle is sufficient to 'knock' an atom from the surface of the material. The incident particle may not have enough energy to displace a lattice atom, initiating a damage cascade, and so this process can happen at lower energy. It is this process that gives rise to problems within a fusion core, as it could poison a plasma and prevent fusion from occurring. To minimise the potential for poisoning of the plasma, Be is chosen as the first barrier.

This choice is made for multiple reasons, not the least of which are Be's low atomic number, high thermal conductivity and compatibility with the plasma, i.e. the plasma can accommodate a significant amount of Be before fusion collapses. However, these need to be set against the low melting point and significant damage from neutrons. In many ways, Be is not ideal, but as the key criterion in a fusion reactor is a stable plasma that can induce fusion, the material challenges need to be overcome. Put simply, if we cannot induce fusion to generate energy, then regardless of how well the reactor is built, it will have failed.

So far we have spoken about the sputtering of the Be into the plasma, but are there other damage processes that we need to understand?

8.8 Gas bubble formation

With the primary product in D,T fusion being a ^{4}He nucleus and a neutron, each with high energy, it would be expected that one possible form of damage would be from He gas build-up, and indeed this is the case. However, the methods for the build-up involve two separates processes, the first is implantation of He directly from the plasma, while the second arises from transmutation of Be.

8.8.1 Direct implantation of He from plasma

This process follows the methodology that we discussed in chapter 2, and can be modelled in the same manner. The key difference, however, is the energy of the incident particle; when the particle hits the surface, the energy can drop from MeV down to eV. How is this possible?

The best way to answer this question is to think about how people behave in crowds. When there are few people around, it is possible to move as quickly as you like. Once the number of people increases, however, everybody slows until they are moving at about the same speed. This basic principle occurs with particles moving in a plasma/gas, in this case the high-energy particles formed from the fusion transfer their energy to their surrounding atoms, heating the plasma. As a consequence, any ^{4}He nuclei that impact upon the surface can have a wide range of energies, and form bubbles close to the surface (figure 8.10).

Transmutation

The second mechanism for forming He inside the first wall is more insidious, arising from a (n,2n) reaction between the ejected neutrons and the Be facing the plasma. In some contexts, this reaction is advantageous, but in this one it is not, and it causes problems.

$$^{9}_{4}\mathrm{Be} + ^{1}_{0}\mathrm{n} \rightarrow 2^{4}_{2}\mathrm{He} + 2^{1}_{0}\mathrm{n}. \tag{8.2}$$

This reaction can happen in the first few atomic layers of the Be coating, or at deeper depths, giving rise to volumes of increased stress/strain, which can lead to crack initiation and the failure of the material. As with fission gas formation, the atoms are often mobile and begin to agglomerate within the structures, forming bubbles.

So, to answer the question posed earlier, Be is used as it matches the necessary criteria for ITER, i.e. it provides a barrier suitable for the effects of fusion, while at the same time not inducing further problems within the plasma. But how will these changes modify the properties of the first wall, and are they likely to prevent long-term use within the core?

Fundamentally, the effects of damage in the first wall are those that would be expected, as discussed in chapter 2, i.e. embrittlement, induced stress/strain and gas bubble formation. These changes are not restricted to the first wall, in many respects they will be experienced by the divertor, where, given the nature of the divertor, they are likely to magnified.

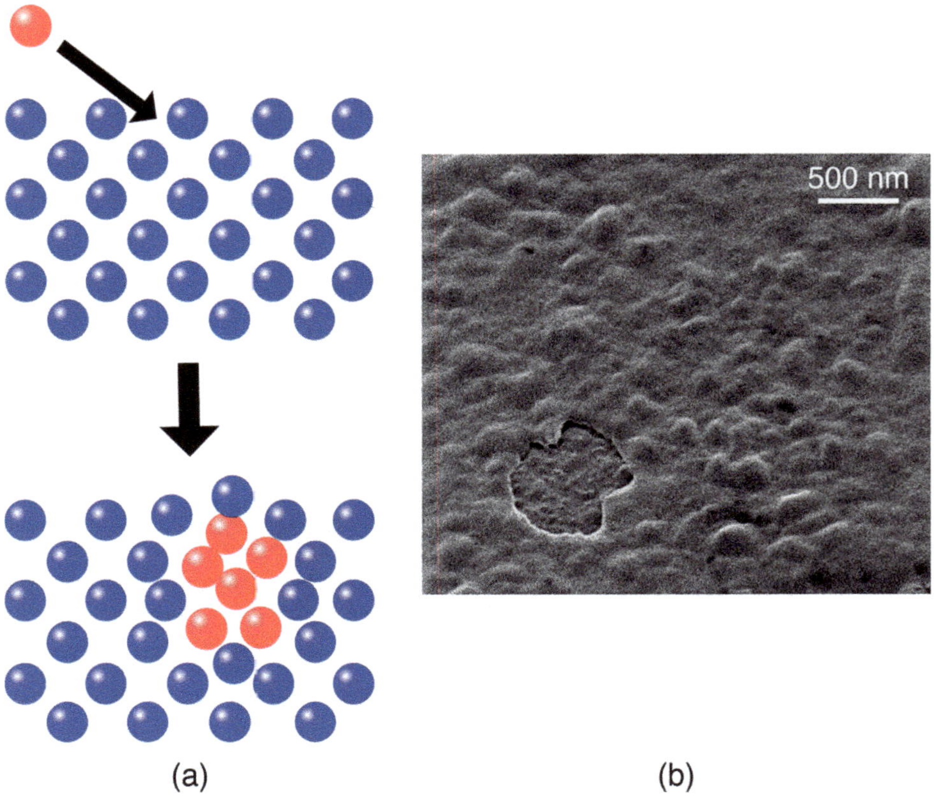

Figure 8.10. (a) Schematic of He bubbles formed on the surface from implantation. (b) Example surface from irradiation of a $CaTiO_3$ ceramic with low-energy He ions within a helium ion microscope [9].

8.9 The divertor

The purpose of the divertor is to channel the gaseous waste product from fusion, in this case $^4He^+$, out from the plasma using magnetic fields. Using figure 8.11 as a guide, as the plasma is contained within the field, the gases are diverted down into the divertor assembly, where they are removed from the plasma. It is a peculiarity of the ITER design that the divertor will experience more extreme conditions than those for the first wall in terms of both the temperatures expected and the radiation damage induced. Luckily, there is greater freedom of choice regarding the materials that can be used in this part of the design, and at least for ITER tungsten is being used. What are the conditions that the divertor is expected to experience?

As with the first wall, the divertor is expected to experience radiation damage, both internal and sputter induced, gas bubble formation and, as the plasma is in contact with the first wall, high temperatures. The divertor is expected to face a heat load of 10 MW m^{-2} [4], which is higher than the heat loads experienced by re-entering space craft. The divertor assembly, shown in figure 8.12, comprises multiple components, with the plasma-facing plates designed to be the barrier

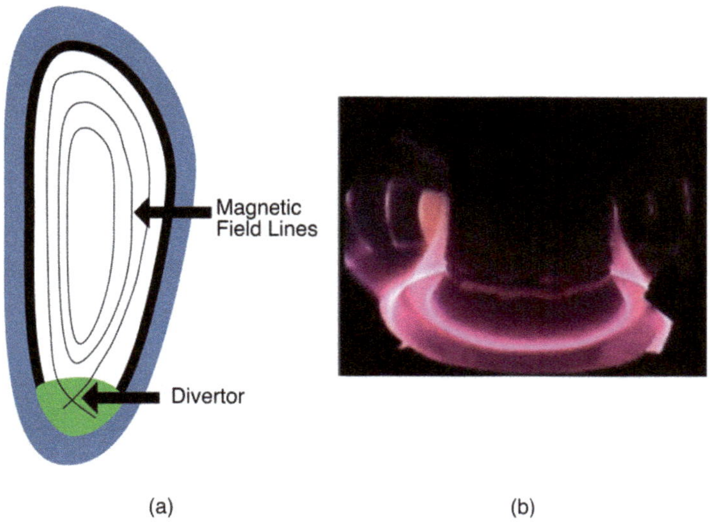

Figure 8.11. (a) Schematic of the containment and diversion of the He into the divertor using magnetic field. (b) Example of a plasma within JET from [3].

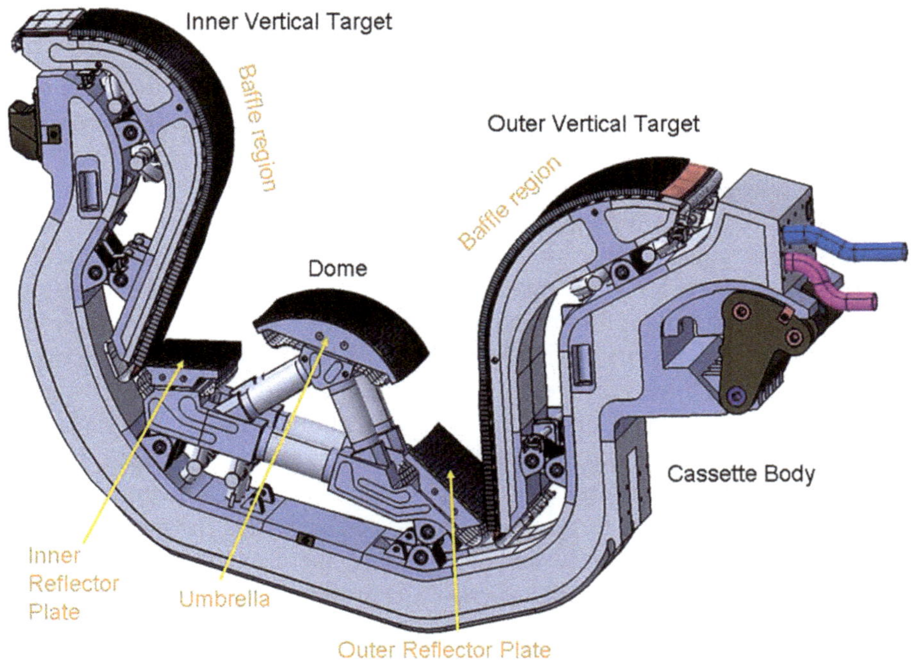

Figure 8.12. Outline of the divertor assembly proposed for use within ITER [4].

between the structural components and the plasma—for ITER they are composed of tungsten. So, can we predict how the tungsten will behave?

The effects of damage can be examined, predominantly through ion implantation using He accelerated to a range of energies in an ion beam, or via a plasma within a

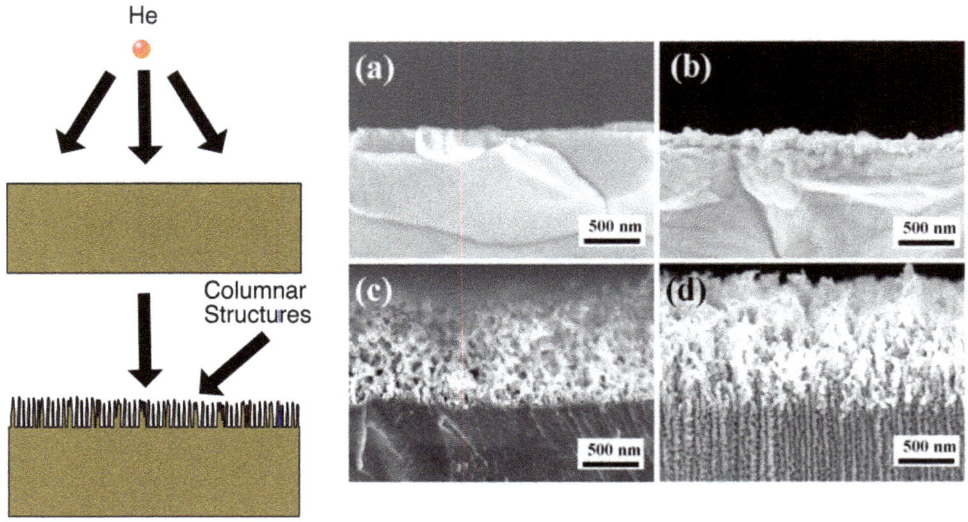

Figure 8.13. (left) Schematic of fuzzy tungsten formation on the surface of the divertor plates. (right) Examples of a columnar structure forming on the surface of tungsten irradiated with He^+ ions at (a) 12 eV, (b) 30 eV, (c) 70 eV and (d) 120 eV. He^+ irradiations were performed at He^+ flux of 1.6×10^{22} m^{-2} s and He^+ fluence of 1.0×10^{26} m^{-2}. Specimen temperatures were (a) 1350 K, (b) 1360 K, (c) 1430 K and (d) 1490 K from [10].

plasma chamber, with the results obtained so far showing some interesting results. These results have revealed challenges that need to be addressed or overcome for future reactors such as DEMO, if tungsten is to be used.

Looking back at the first wall and the possibility of sputtering, a similar process is expected in the divertor, and since we are not likely to poison the plasma here, it should have less interaction with the fusion process. However, there still is a challenge that needs to be overcome, as the effect of sputtering is potentially more pronounced, and can lead to failure in the material. Since the He^+ being diverted has effectively lost most of its energy, i.e. been thermalised, it will no longer be expected to have energies in the MeV range, they are more likely to be in the eV range. The He^+ will still have sufficient energy to form gas bubbles within the surface, and might have sufficient energy to sputter some W atoms from the surface, but it will probably not have enough energy to move them far. As a result, the tungsten surface begins to be become 'fuzzy', forming pillars on the surface, as shown in figure 8.13.

This phenomenon is not fully understood, and there are multiple possibilities available to explain it; however, regardless of the mechanism for their formation and growth, it is the consequences of their existence that require a solution. Such a change in the structure of the tungsten has implications, and changes the material behaviour. So, what could they be?

If we were to consider this purely from a radiation damage viewpoint, the normal effects of damage would be sufficient, for example, increased hardness, reduced ductility, inducement of stress/strain, etc. For example, W has been found to increase in hardness with increasing levels of damage induced by He implantation [11]. There is, however, a reduction in thermal conductivity near the surface due to damage, which, given the likely increased heat load in DEMO, needs to be more fully understood, and material changes may be be needed.

Given the issues surrounding tungsten and, for that matter, beryllium in the first wall, are there are any other materials that could be used? Some alternatives have been proposed, based on a range of material options, from the ceramic (e.g. C and SiC) through to the metallic (e.g. W and BeW), and even materials that have properties of both, such as MAX phases and others. Each of these materials has advantages and disadvantages, and this is the focus of ongoing development.

Using SiC as an example of behaviour within a fusion core, and predicting some likely changes from induced damage, we can expect the following.

i) Lowering of thermal conductivity—high-purity SiC can have high thermal conductivity (~400 W m^{-1} K), but with induced damage this can drop dramatically [12].
ii) Expansion due to damage—as we discussed in chapter 6 when talking about TRISO fuel, which contains SiC, there is an expansion, which in this case is accommodated by the surrounding C layers. Here there would be expansion into the plasma, providing more surface to interact with the evolved neutrons/$^4He^+$.
iii) Cracking—with a ceramic, which lacks the ductility of a metal, the inducement of stress/strain within the lattice will lead to increased cracking.

So far, we have discussed the basic design, and plasma-facing components, but we have not yet discussed how they will be used to generate more fuel and electrical energy.

8.10 Breeding and heat generation

As we outlined earlier, there are two products produced from D,T fusion, a ^{4}He nucleus and a neutron. So far, we have discussed the nature of the damage they induce and their impact on reactor materials. However, in future reactor designs the neutrons produced will have two uses:

(i) transfer of energy to generate electricity;
(ii) breeding of further tritium for fusion.

Before we discuss the breeding question, how can the neutrons be used to generate electricity?

Essentially, they would be used in a manner similar to fission reactors; the heat they generate when pacing through, or impacting on the first wall, will be transferred to a medium and then used to drive turbines. In ITER, the first wall/reactor vessel will be water cooled, but there are no plans to convert the energy to electricity; this will be addressed with DEMO. From a schematic viewpoint, similar challenges to those experienced in fission reactors will present themselves, for example, if water is used to transfer the energy, the effects of induced radiolysis will need to be accommodated, i.e. increased corrosion.

What is more of a challenge is the generation of fuel for the reactor. This requires materials that not only resist damage from neutrons, but also allow their transport, and this will need to be coupled to a method for generating ^{3}H. So, how might we achieve this?

8.11 Tritium breeding

There are multiple methods available for tritium breeding, but they follow the same fundamental process, neutron capture, they just differ in their implementation and product. The easiest way to breed tritium is through the transmutation of ^{6}Li or ^{7}Li.

$$^{6}_{3}\text{Li} + ^{1}_{0}\text{n} \rightarrow ^{4}_{2}\text{He} + ^{3}_{1}\text{H}. \tag{8.3}$$

$$^{7}_{3}\text{Li} + ^{1}_{0}\text{n} \rightarrow ^{4}_{2}\text{He} + ^{3}_{1}\text{H} + ^{1}_{0}\text{n}. \tag{8.4}$$

In each case, the products are ^{4}He and ^{3}H, with the addition of a ^{1}n when using ^{7}Li. There is one issue to consider, the rate at which the ^{3}H can be formed is limited. If we assume that all of the neutrons that enter the breeder can be used for transmutation, we will still not be able to generate sufficient ^{3}H. Why?

Put simply, the transport of neutrons through materials is not 100% efficient, and neutrons will be lost. We therefore need to increase the efficiency of the process. How might we do this?

So, having increased the efficiency of the use of the available neutrons, how could we breed ^{3}H? As with nuclear fuel, there are two main types, solid and liquid breeding, each bringing its own challenges and advantages.

8.11.1 Solid breeding

This approach uses ceramics, such as Li_2TiO_3, formed as small porous pellets, which are themselves contained within a solid vessel lined with Be, and cooled by He gas (figure 8.14). As the transmutation proceeds, the formed ^{3}H is then transferred out with the He gas and separated.

8.11.2 Liquid breeding

This approach can use one of two methods, both of which we have come across before, when discussing fission reactors. One method utilises a Pb–Li alloy, similar to the GenIV LFR, while the other uses a molten salt eutectic, similar to the GenIV MSR.

8.11.3 Pb–Li eutectic

This foregoes the use of Be as a neutron multiplier, as ^{204}Pb, ^{206}Pb, ^{207}Pb, and ^{208}Pb can all undergo (n,2n) reactions, therefore providing the multiplier internally, greatly increasing its efficiency.

$$^{204}_{82}\text{Pb} + ^{1}_{0}\text{n} \rightarrow ^{203}_{82}\text{Pb} + 2^{1}_{0}\text{n}. \tag{8.5}$$

As with solid breeding, the approach would use He as a cooling gas, but would use a flowing Pb–17.5Li circuit, allowing easy replenishment of Li and extraction of ^{3}H. A schematic is shown in figure 8.15.

8.11.4 Molten salt—FLiNaBe

This method uses a similar approach to that used in molten salt reactors, but in this context it will be used to generate ^{3}H from $^{6}Li/^{7}Li$ with an internal neutron

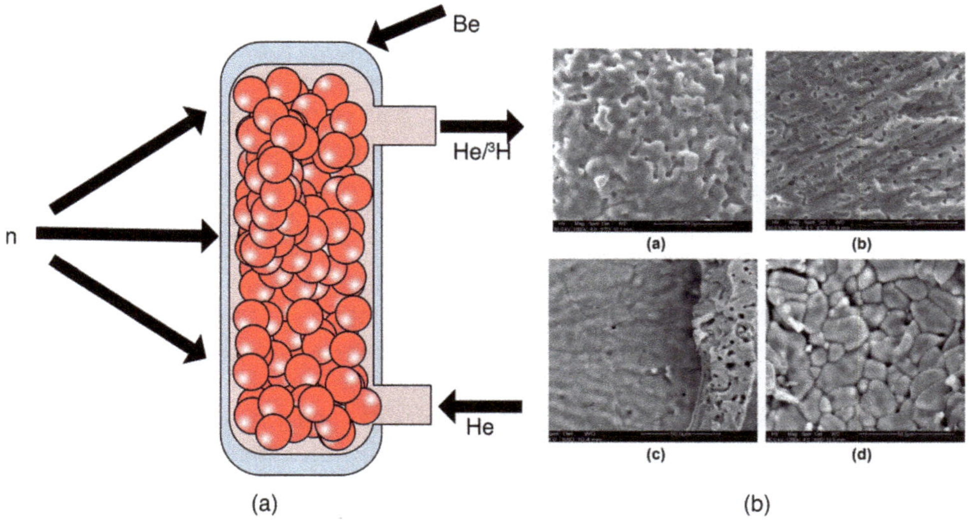

Figure 8.14. (a) Schematic of a solid ^{3}H breeding device. (b) Example of Li_2TiO_3 beads, showing different fracture surfaces and the pores within the lattice from [13].

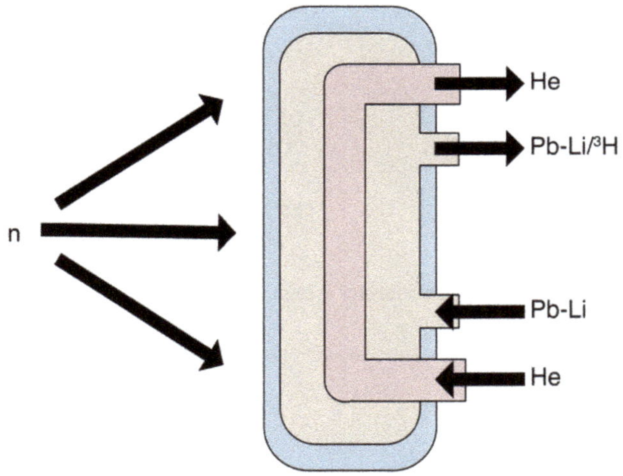

Figure 8.15. Schematic of the Pb–Li liquid ^{3}H breeder.

multiplier, ^{9}Be. This design has similar advantages to the Pb–Li eutectic, and allows continuous replenishment and extraction of the ^{3}H.

8.11.5 Comparison of methods

All of the methods proposed for breeding have the ability to transfer the heat away from the core, which can then be used to generate electricity. There are advantages and disadvantages to each method, for example, when using solid breeding material, the transference of the formed ^{3}H out of the solid will need to be perfected. For example, what is the desirable microstructure, and will it be able to accommodate the inherent radiation damage induced? Whereas in the liquid cases, the effects of corrosion, enhanced by gamma radiolysis, similar to the effects we discussed in chapter 6, will need to be accommodated. However, since these methods are similar to those proposed in GenIV fission designs, this should be achievable. However, one question still arises—after we have formed ^3H, how is it introduced into the core?

We can introduce ^3H into the core using two routes. One is diffusion, but this is not the most efficient method. A more desirable method, which implants $^2H/^3H$ directly, ready for fusion, is to freeze the gas into a pellet and 'fire' this into the plasma, as shown in figure 8.16.

This approach would freeze the purified ^3H and ^2H into cylinders (~6 mm diameter and 6 mm length) and fire them into the plasma at ~20 Hz, with a speed of ~360 m s^{-1}, where they would then volatilise and be ready to undergo fusion in the ideal location.

We have discussed the challenges for the plasma-facing first wall, the divertor, tritium breeding and introduction, and the cooling of materials—can we make comparisons between GenIV fission and fusion reactors?

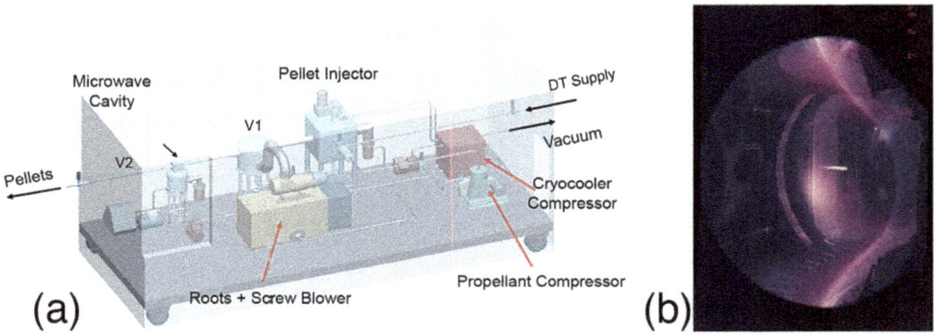

Figure 8.16. (a) Design of $^2H/^3H$ pellet transfer system from [3]. (b) Pellet of $^2H/^3H$ in JET from [4].

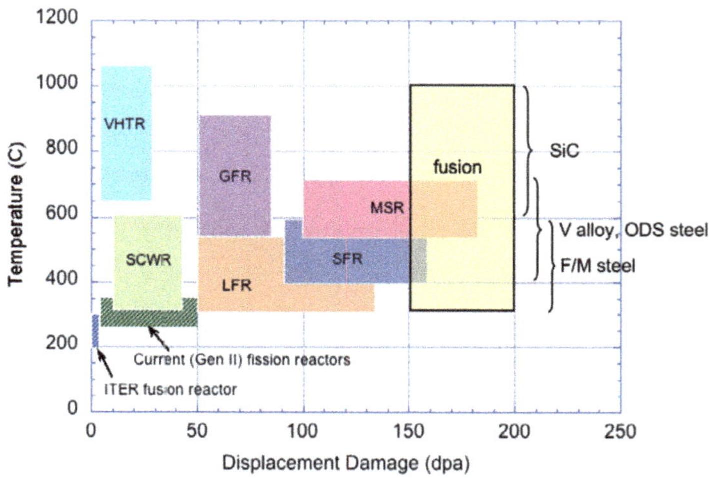

Figure 8.17. Outline of the regions of overlap in fusion/fission materials development. Reproduced with permission from [14].

8.12 Challenges in fission and fusion

The development of materials for advanced nuclear technology no longer divides into purely fusion- or fission-based research, the two are linked together, as in many ways the challenges that impact on fission reactors are similar to and directly comparable with those for fusion reactors.

If we use figure 8.17 to make a comparison, in both fission and fusion reactors high levels of damage and temperature are predicted, and these are comparable. As such, there are common material challenges that need to be overcome, and some materials have potential in both technologies. For example, SiC has been proposed for use in the VHTR GenIV reactor, within the fuel and as a structural component, while in fusion it has been proposed for the divertor. In many ways, the methodology applied to fission reactor materials design is also applicable to fusion reactor materials. For example, it often involves a balance of desirable against undesirable properties, or, in reality, what we need and what we can live with.

The ITER design is only the first stage in the process of developing commercial nuclear fusion, the next two steps will be DEMO and PROTO [4]. For example, in DEMO, the breeding of tritium will be developed, coupled with a means by which the energy can be extracted and converted to electricity. One of the key challenges in the design of DEMO is to understand the effects of damage within a fusion core; it is expected to experience higher levels of damage from high-energy neutrons and $^4He^+$ ions. This level of damage is not easy to model, and with there being a scarcity of facilities able to irradiate with 14.1 MeV neutrons, it is difficult to experimentally assess materials, but with enhancements to facilities, it will be achievable. The final stage in the development, PROTO, will focus on the development of a commercial fusion power plant.

In this chapter, we have discussed the process of nuclear fusion, how it can be achieved terrestrially and what material challenges still need to be overcome. In the next chapter, we shall return to the world of fission reactors and discuss what happens when they go wrong, what has been learnt from past cases and how future designs have been modified.

References

[1] Wieser M E *et al* 2013 *Pure Appl. Chem.* **85** 1047–78

[2] Chadwick M B *et al* 2006 ENDF/B-VII.0: next generation evaluated nuclear data library for nuclear science and technology *Nuclear Data Sheets* **107** 2931–3060

Chadwick M B *et al* 2011 ENDF/B-VII.1 nuclear data for science and technology: cross sections, covariances, fission product yields and decay data *Nuclear Data Sheets* **112** 2887–996

[3] Culham Center for Fusion Energy www.ccfe.ac.uk

[4] ITER www.iter.org

[5] Max Planck Institut fur Plasmaphysik www.ipp.mge.de/16900/w7x

[6] University of Wisconsin Madison www.hsx.wisc.edu

[7] Yi X, Jenkins M L, Hattar K, Edmondson P D and Robert S G 2015 *Acta Materialia* **92** 163–77

[8] Voitsenya V S, Balden M, Bardamid A F, Belyaeva, A I, Bondarenko V N, Skoryk O O, Shtan A F, Solodovchenko S I, Sterligov V A and Tyburska-Püschel B 2014 *J. Nucl. Mater.* **453** 60–65
[9] Lawson M S and Gandy A S 2016 private communication
[10] Liu L, Liu D, Hong Y, Fan H, Ni W, Yang Q, Bi Z, Benstetter G and Li S 2016 *J. Nucl. Mater.* **471** 1–7
[11] Armstrong D E J, Edmondson P D and Roberts S G 2013 *Appl. Phys. Lett.* **102** 251901
[12] Katoh Y, Snead L, Szlufarska I and Weber W J 2012 *Curr. Opin. Solid State Mater. Sci.* **16** 143–52
[13] Vittal Rao T V, Bamankar Y R, Mukherjee S K and Aggarwal S K 2012 *J. Nucl. Mater.* **426** 102–8
[14] Zinkle S and Busby J T 2009 *Mater. Today* **12** 12–19

IOP Publishing

Nuclear Materials Science

Karl Whittle

Chapter 9

Mistakes made and lessons learnt

We have discussed the basis for nuclear reactors, the materials challenges they face and the outlook going forward. However, as in other areas, nuclear reactor technologies have also had to learn from negative experiences, with a focus on how to prevent such events occurring again. This is not unique to nuclear reactors, for example, air accidents are investigated and lessons are learnt from them, as in the case of the Comet jet liner, where stresses around the windows led to failure.

Luckily for nuclear technologies, there have been few major incidents. In this chapter, we shall examine four examples that highlighted the challenges reactor materials face, and discuss the lessons learnt. The four examples are:

(i) Windscale—Pile 1;
(ii) Three Mile Island—Reactor 2 (TMI-2);
(iii) Chernobyl—Reactor 4;
(iv) Fukushima—Daiichi complex.

Each of these incidents highlighted a materials challenge/failure, and led to modifications in future designs, or preventive mechanisms being put in place.

9.1 Windscale—Pile 1

During the race for nuclear weaponry, the UK designed and built a pair of reactors for the breeding of ^{239}Pu from ^{238}U, on the west coast of Cumbria, near a village called Seascale. This site became known as Windscale, and is now known as Sellafield. The reactors were built from 1947, and Pile 1 began operation in 1950, with Pile 2 becoming operational in 1951.

The design of the reactor was similar to those of X10 and Hanford B, as discussed in chapter 5, but had a modified cooling system. Whereas the Hanford reactor used water to cool its core, the Windscale piles used forced air cooling, similar to X10. The method of operation was similar, with packets of U, contained within Al cladding, pushed from one face of the core into the core, and eventually out the

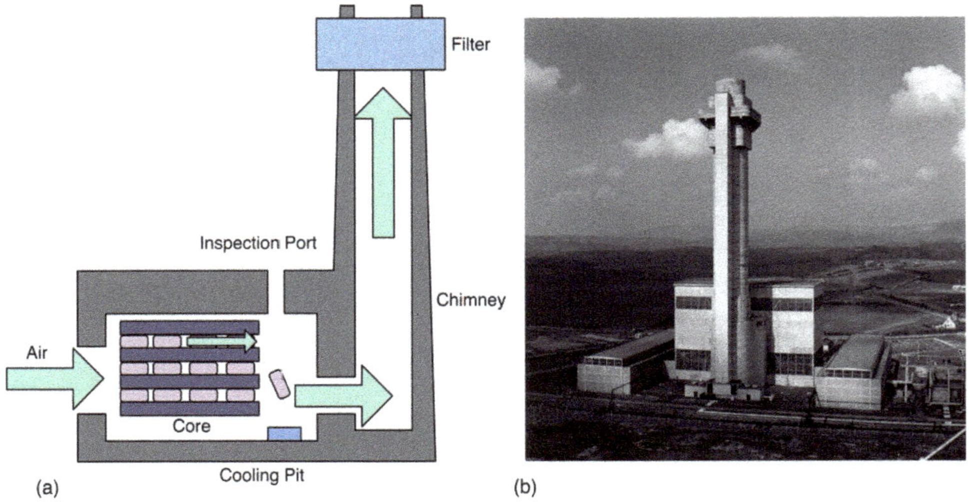

Figure 9.1. (a) Schematic of the reactor, indicating the flow of air, the cooling pit for discharged capsules, the inspection points and the filter place in the chimney. (b) Photograph of the pile after completion from [1].

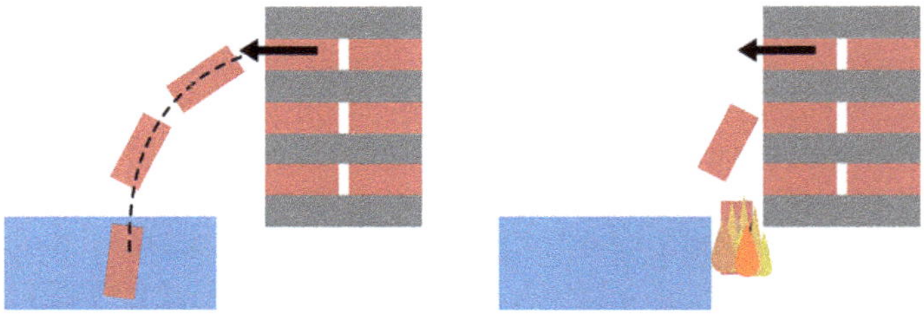

Figure 9.2. Schematic for normal operation within Windscale Pile 1, and for when a can failed to hit the cooling pond, see also animation.

other side. The reactor contained B-steel control rods to moderate the reaction. A schematic of the reactor design is shown in figure 9.1, together with a photograph of the reactor in operation.

During construction, the question was postulated, 'what would happen if one of the fuel cartridges being pushed cracked or missed the channel?'

As you can imagine, a cracked can of Al-clad U metal at high temperature within an oxygen atmosphere would be problematic, and highly likely to catch fire, as shown in figure 9.2. Since the core was being forced air cooled, with a chimney, the programme director, John Cockcroft, insisted that a filter be placed in the chimneys to prevent the release of radioactive material. At the time it was considered by many engineers to be a waste of time and the filters became known as 'Cockroft's folly'.

During the early stages of operation it was decided that the Al-based cladding was too neutron-absorbent, reducing the efficiency of the transmutation from ^{238}U to ^{239}Pu (chapter 3). A decision was made to reduce the cladding thickness, thus increasing the rate of formation. One other issue arose during the early stages of operation, concerning the core itself, and it was an issue that needed to be addressed quickly, as if not rectified it could lead to a core failure, or worse. The issue related to the neutron damage occurring with the core, and the build-up of Wigner energy.

9.1.1 Wigner energy

Eugene Wigner (1902–1995), Nobel prize winner in 1964 and one of the scientists in the Manhattan project, is credited with two discoveries, both ironically leading to issues with nuclear reactors. One concerns an isotope of xenon, which we will discuss with relation to Chernobyl, while the other relates to the effects of neutron damage within graphite [2].

When we discussed the effects of radiation damage in chapter 2, we outlined the possible effects from cascade damage. Wigner energy primarily concerns those neutrons/atoms that occupy interstitial positions after a cascade/damage. These non-ideal locations retain potential energy, which, if there are few, the lattice can accommodate. When the number becomes sufficiently large, these atoms raise the potential energy of the lattice, and can release the energy spontaneously. This is particularly a problem for graphite, but why?

9.1.2 Graphite

Graphite, as we all know, is one of the polymorphs of carbon, and is based on carbon atoms arranged hexagonally in sheets, sharing electron density with adjacent sheets. As such, it can be soft, and has multiple uses, for example, within a reactor core. If we damage such a structure, then, as previously discussed, there can be damage cascades, which in the case of graphite can lead to the structure being distorted away from the ideal flat sheets. This distortion strains the lattice, which would ideally relax away, and not build-up. However, at low temperatures it does not anneal out, and this leftover energy builds up (Wigner energy). If this is not ameliorated, then the ultimate outcome could be failure within the core. So, how can we prevent the build-up of Wigner energy?

One way to remove the problem is to run the reactor at high enough temperatures for annealing to occur, such as within an AGR. However, at the Windscale piles this effect was only just beginning to be understood, and since they ran at low temperatures, the effect would have increased continuously. The method chosen at Windscale, and other reactors at the time, was to raise the temperature of the core and allow the Wigner energy to release; for the Windscale piles, 250° C was chosen (figure 9.3). The reactor would be heated to this temperature, and the Wigner energy would be released as heat, allowing the reactor to anneal and return back to a less stressed state.

During the operation of the Windscale piles, the nature of the irradiations changed from pure breeding of ^{239}Pu from ^{238}U, to include breeding of ^{3}H from ^{6}Li/^{7}Li (see chapter 8). As a consequence, the nature of the cladding changed, with it now being thinner than initially designed. Between 1953 and 1957 the Wigner energy was released eight times for Windscale Pile 1.

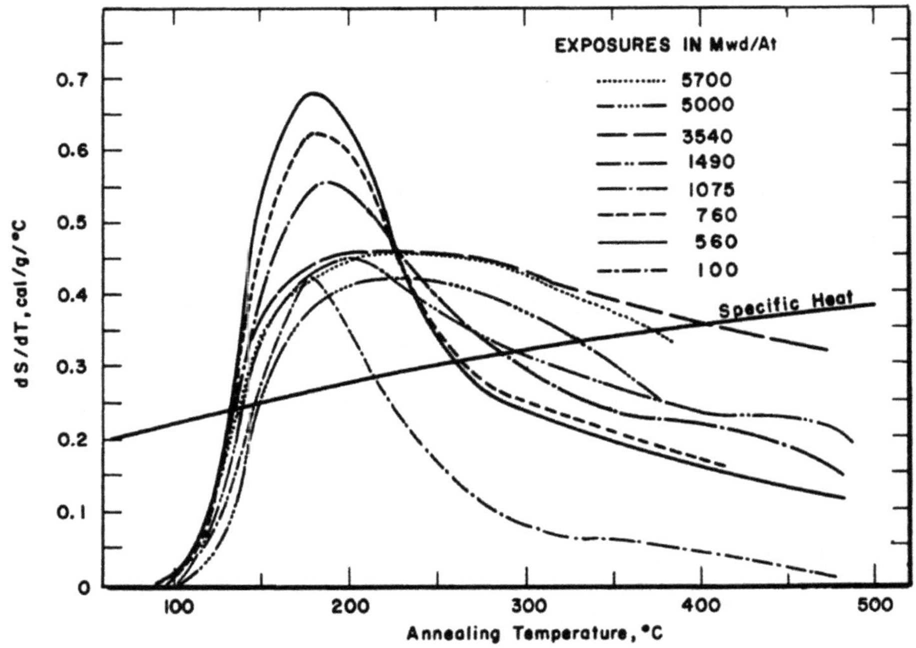

Figure 9.3. Energy release rate against temperature for graphite, with differing degrees of irradiation [3].

9.1.3 Timeline of incident

The timeline reported here is taken from multiple sources, which sometimes contradict each other. However, the timeline is the generally accepted version of events [4–6].

7 October 1957

11:45. It was noticed that the reactor was heating up, and since this could indicate an uncontrolled release of Wigner energy, the reactor was closed and a release of Wigner energy authorised. The core began to heat up.

8 October 1957

01:00. It was noted that the core was heating abnormally, with most thermocouples varying from 50° C–80° C, however, one thermocouple (channel 2053) was reading a temperature of 210° C, indicative of Wigner energy release. The energy release was considered to be localised in the region around 2053.

11:05. The reactor was not rising sufficiently, and the pile was restarted, keeping the fuel at ~330° C, with heating continuing until **19:35**. It was noted that the reactor was 'touchy' at around 330° C, fluctuating in temperature.

9 October 1957

The temperature continued to rise to 415° C, when the cooling fans were switched on and the dampers opened. It was unfortunate that at this time Pile 2 recorded an increase in radioactivity, indicative of a fuel canister breach, so attention was diverted there.

10 October 1957
05:40. The radiation monitors in the chimney reported increased levels of radioactivity, and it became apparent that the pile was in trouble. By **13:30**, the monitors were continuing to show high levels, with the reactor core still at ~420° C and showing no signs of dropping in temperature. To try to reduce the temperatures at ~**14:30**, the fans were turned back on and air forced through the core. The core was examined, from the side where fresh material was introduced and the channel, 2053, recording the high temperature was found to be glowing orange, indicative of the core being on fire.

To minimise the chance of further contamination, the fuel billets around this channel were pushed through, and out of the core. The workers were protected by the cooling air providing a barrier between them and the fire. It was at this point that a visual inspection of the core, through one of the viewing ports above the core (figure 9.1) was made, and it was confirmed that the core was indeed on fire. The concern now became that at these temperatures, the concrete support structure could become unstable. It was clear that the fire had to be extinguished, and quickly, but the concern was how to do this, as water at this temperature could split into hydrogen and oxygen, and what was a fire could become an inferno.

11 October 1957
In the first attempt to extinguish the fire, CO_2 was brought from the nearby Calder Hall reactor complex and the core was flooded with a barrier gas. However, this did not work and the fire continued; the only chance left now was to extinguish the fire by pumping water directly into the core. The decision was made in conjunction with the local authorities, and the ground was set for further action, if it proved necessary.

A major problem was discovered, as water was never expected to be used, there was no set-up that could be used. So a jury-rigged injection system based on scaffolding tubes was used to pump water directly into the core.

09:00. The water was switched on, and as luck would have it, the reactor did not explode. However, from the viewing port it could be seen that the fire was still raging, and not showing any sign of abating. The fans were still on, passing air through the core, from when the fuel was being pushed through, and this was providing one of the necessary ingredients for combustion, oxygen to fuel the fire. At **10:00** the fans were turned off and the fire began to subside until it eventually stopped as the oxygen ran out.

12 October 1957
After 24 h of pumping, the water was turned off, and by the afternoon, the pile was cold.

9.1.4 Aftermath

Close examination of the core discovered that ~10 tonnes had melted, and this melted material had formed a solid crust, composed of oxides that did not travel very far. However, the concern was now the volatiles that would have been released, such as ^{131}I, ^{137}Cs, and ^{133}Xe. Whilst these were indeed released, the levels released were found to be much lower than anticipated, as what had once been described as Cockcroft's Folly had filtered significant amounts—it could no longer be considered a folly in any way. The filters, while not being able to remove everything, removed significant amounts of material, and prevented further contamination. Since there was a release of radioactive isotopes, such as ^{131}I and ^{137}Cs, which can enter the biosphere, the local farmers were ordered to destroy their milk for a significant time afterwards.

As a direct result of the incident in Pile 1, the second reactor, Pile 2, was shut down on 17 October 1957, and neither reactor was ever operational again. The use of air-cooled reactors was stopped worldwide.

9.1.5 What caused the fire?

The cause of the reactor fire was thought to be a failure in one the fuel billets, and there are differing explanations for the cause of this. One is that the failure occurred as a direct consequence of the second Wigner energy release cycle. An alternative is that there was a simple failure in the cladding, allowing the lithium within to react with the carbon and aluminium, igniting the surrounding material. However, given the secrecy of the times, it is unlikely we will ever know for sure what started the fire, we just know that it did start.

9.1.6 Key outcomes from the fire

Other than the immediate outcomes, such as the closure of Pile 2, the incident led to changes in reactor material use and the design of safety systems. For example, the filtering of air from reactor systems, which before the fire had been thought unnecessary, was now required. This led, in turn, to a change in the role of safety planning, and a move towards a 'what could happen?' scenario, as opposed to a 'what is likely to go wrong' approach.

One of the key issues, the release of Wigner energy, raised a potential issue with using graphite as a moderator/structural component within a core. However, this problem does not arise if the reactor is run at high temperature, generally above 250° C, and since the next generation of reactors, such as the MAGNOX at nearby Calder Hall, ran at 360° C, this problem was no longer present.

As the incident arose from a failure in cladding, the challenge became to make cladding mechanically robust, with a minimal failure rate. Since the cladding at the Windscale piles was Al-based, and the majority of reactors had moved to Zr-based cladding for LWR, the challenges were different. It was one of these challenges however, that arose at Three Mile Island.

9.2 Three Mile Island—Reactor 2

It is often said that truth can be stranger than fiction, and that fiction can often predict fact. Such a statement could be made about Three Mile Island and the China syndrome, which has become a name associated with a reactor meltdown.

9.2.1 The China syndrome

The film *The China Syndrome* used as its premise a problem with a fictional reactor located in California leading to a potential core meltdown. The title refers to the idea that a core meltdown would cause the core to melt into the Earth and continue to travel 'all the way to China', which is highly fictional. However, as with a lot of phrases, it has taken hold when talking about nuclear energy and is used erroneously. In the film, the problem arises from a leak within the cooling system leading to the coolant being dangerously low, in essence leading to a loss of coolant accident (LOCA). The film was released on 16 March 1979.

9.2.2 Three Mile Island nuclear reactor complex

The Three Mile Island power plant was located near Harrisburg, Pennsylvania, and comprised two Babcock and Wilcox PWR reactors [7], as shown in figure 9.4. The reactors were:

(i) TMI-1 with an output of 852 MWe, and was commissioned in 1974.
(ii) TMI-2 with an output of 906 MWe, and was commissioned in 1978.

9.2.3 Timeline of incident

The timeline reported here is the generally accepted version of events [8].

27 March 1979
There was a malfunction in one of the filters within the secondary cooling circuit, TMI-2; this was rectified and the reactor continued operating.

28 March 1979 (12 days after The China Syndrome *was released)*
04:00. Workers were undertaking maintenance on the secondary cooling circuit when it was noticed that the pumps had stopped. The pumps switched to the back-ups.

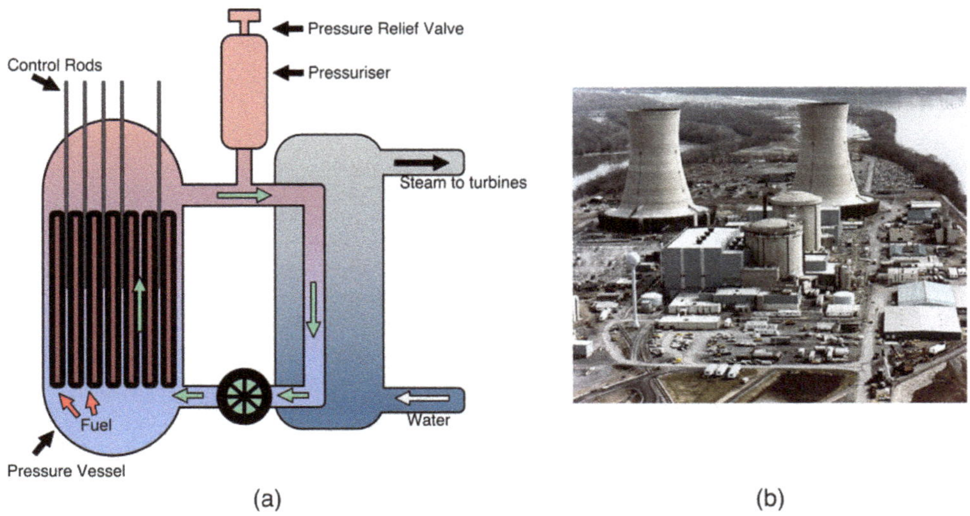

Figure 9.4. (a) Schematic of a PWR. (b) The Three Mile Island reactor complex from [7].

However, there was no water flowing through the circuit. The turbines, sensing there was no water flowing, shut down, and the reactor SCRAMed. At this time there was no cooling water in the secondary cooling circuit, and the primary circuit began to overheat. To alleviate this, the pressure relief valve was opened; it should have shut down after 10 s, and the control room read that the valve had closed.

05:00. The primary cooling circuit pumps began to vibrate; they were now pumping a mixture of steam and water. This was recognised at **06:22**, when it was discovered that the valve had not closed and was still open. The valve was now closed, stemming the flow of steam out of the core, and now the problem was the superheated steam being located within the reactor, with nowhere to go; it was blocking the flow of liquid water into the core. The situation had now become a loss of coolant accident (LOCA), and the fuel started to overheat and subsequently melt.

07:45. Notification of the nuclear regulatory commission, the state governor and the White House began. The issue now was that as the fuel in the core was not being cooled or covered by water, it began melt. At **10:30** radiation from the core was beginning to be detected in the control room and the people present had to start wearing respirators. Throughout the day, work continued to restart the pumps, and at **19:50** one pump had started and was beginning to push water back into the system. However, by this time, the damage was already done, hydrogen had already been formed from reaction between the Zircaloy cladding and the superheated steam, and the pressure was increasing.

29 March 1979

There was increasing concern regarding hydrogen build-up within the core, and the potential for a core rupture, and this was alleviated by opening the pressure release valve that had previously opened periodically. This process continued for the next few days, until the risk subsided. An animation of the events at TMI-2 are shown in figure 9.5.

Figure 9.5. Animation of the stages during the LOCA at TMI-2. Reproduced with permission from [9].

27 April 1979
The reactor was now below 100° C, at ambient pressure and considered to be in safe mode.

During the incident, TMI-1, the adjacent reactor, continued to run normally, and was shut down, reaching safe mode on 7 April. The reactor continues to operate, and was permitted to operate until 2034 in 2009 [7, 9].

9.2.4 Aftermath

As with the Windscale Pile 1 fire, the core was later opened (July 1984) and examined closely. It was discovered that ~45% of the core had melted (62 tonnes), with 19 tonnes at the bottom of the core. It was also determined that while there was a risk of the core melting through the pressure vessel and containment, in this case the metallic elements, mainly uranium and zirconium, had formed oxides that provided a barrier. A schematic of the core outcome is shown in figure 9.6.

Removal of the fuel began in 1985 and was completed in 1990, while shipment of the core debris began in July 1986. The reactor was then effectively sealed, with further decommissioning being delayed until the decommissioning of TMI-1 began, allowing cost savings to be made.

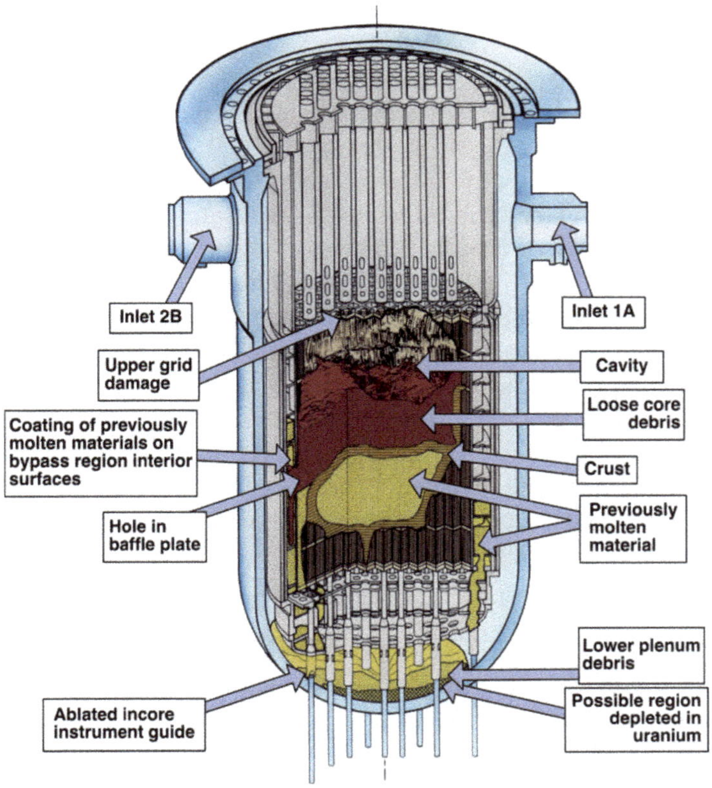

Figure 9.6. Schematic of the end TMI-2 core after the LOCA from [8].

9.2.5 What caused the initial problem that led to the incident?

During maintenance on 27 March with the pumps in the secondary cooling circuit, in order to clear a blockage the engineers inadvertently introduced water into an air line. As a result, this water prevented the automatic operation of the valves that allowed circulation of cooling water.

9.2.6 Key outcomes from the incident

Multiple lessons were learnt from the incident, ranging from a greater understanding of a LOCA through to a greater appreciation of the technical challenges of running a power reactor with complex operation schemes. The key outcome on the operational side concerned reducing system complexity, and inspired Charles Perrow's 'normal accident theory' [10], where a poorly designed system can lead to problems, as interactions not previously considered interact.

The incident at TMI-2, as with Windscale Pile 1, highlighted the need to change how the safety systems within a reactor operated, and the need to increase system redundancy. The Babcock and Wilcox reactor design of TMI-2 is a GenII one, and the lessons learnt from the TMI-2 incident have been incorporated into GenIII/III+ reactor designs. One question we have not answered yet is, what is a loss of coolant accident?

9.2.7 Loss of coolant accident (LOCA)

Put simply, a LOCA occurs when there is no longer sufficient coolant to cool the reactor, and the system moves from being a stable, predictable system, to one that is predictable, but for entirely the wrong reasons.

In chapters 3 and 4, we discussed the nature of the cladding and fuel used in light water reactors, and how they can interact with the water. As we outlined, there are two concerns with Zr-based cladding: corrosion giving rise to hydrogen formation, and the question of where the hydrogen goes. During normal operation within a PWR, the level of corrosion is monitored and managed. However, during a LOCA, when we longer have water but high temperature steam, the reaction increases dramatically, giving rise to the formation of H_2 gas, and ZrO_2 on the surface of the clad (equation (9.1)). Ironically, the formation of ZrO_2 helps to passivate the surface, and prevent further formation of H_2 gas. However, the ceramic is brittle and begins to fall away from the surface, introducing fresh clad for the steam to react with, but nevertheless it does help to slow the process down. An illustration of the process is shown in figure 9.7.

$$Zr_{(s)} + H_2O_{(l)} \rightarrow ZrH_{x(s)} + ZrO_{2(s)} + H_{2(g)} \tag{9.1}$$

A second equally problematic issue that arises from a loss of coolant, is the loss of cooling. When a reactor is operating, the coolant transfers the heat obtained from the fission event out of the core. We then use this heat to generate electricity through turbines. However, not all the energy released from the fission is immediately transferred to the coolant, for example, since fission is a probabilistic event, there are a range of fission products, some of which are short-lived and decay, and there is still

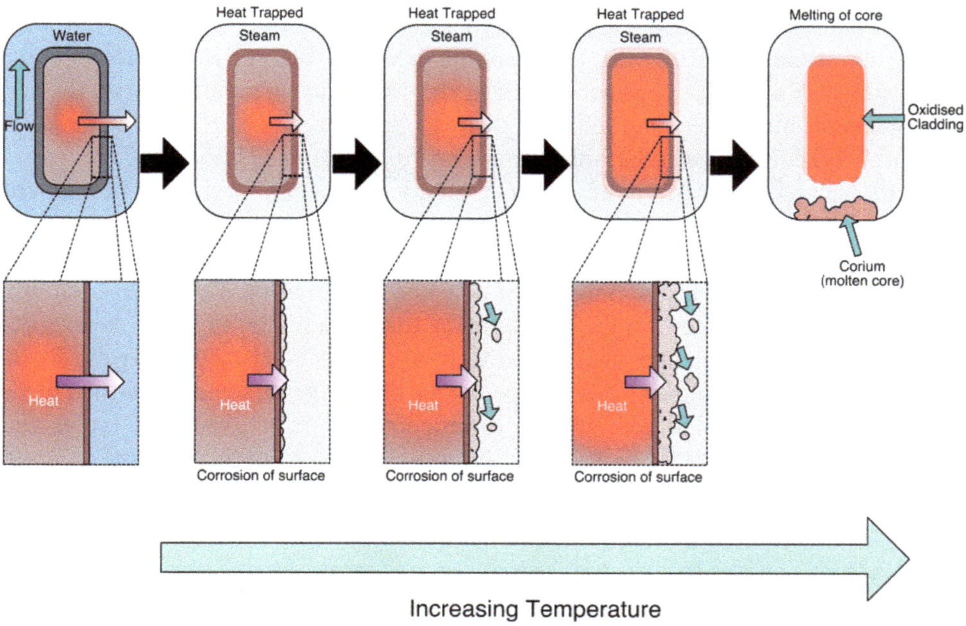

Figure 9.7. Schematic of the stages that cladding undergoes during a LOCA.

energy contained within the system. Hence, even when fission has been stopped, i.e. the control rods have been inserted, there are still short-lived isotopes undergoing decay and releasing energy, i.e. heat.

Now that we have no active cooling in the core, and at best are relying on what is in the core to cool the reactor, we have a problem, the heat has to go somewhere. We compared the different thermal conductivities and melting points of UO_2 and Zr, for example, Zr is a better conductor of heat than UO_2, but UO_2 has a higher melting point than Zr. When there is active cooling, this variance is okay, and ideal, but when we are no longer cooling, we have a mismatch and a problem.

The nature of heat flow is to always flow from hot to cold, this is the zeroth law of thermodynamics, and since in a LOCA we have hot ceramic fuel, with poor thermal conductivity but a high melting point, the heat will transfer through the cladding, which has high thermal conductivity, but a low melting point. This inevitably leads to the clad melting, releasing the fuel into the core, and once this fuel hits the core, it can lead to the basis of the phrase China syndrome—it can melt the bottom of the core. In TMI-2, this molten fuel/clad did not melt the bottom of the core, as the one disadvantage in the thermal conductivity of ceramics in this case helped. It allowed the high-temperature molten fuel to fall onto previously formed Zr–O oxides from the cladding, which had been oxidising. Now, at the bottom of the core, there was a volume of what is now described as 'corium', a mixed bag of components, dependent on the source, but often described as a 'lava' of molten material. It often contains fuel, clad, support structures and multiple other options.

This concern with corium led to the designs we discussed in chapter 5 when talking about GenIII/III+, with, for example, the AP1000 family of reactors having a tank of water above the reactor, which in the event of a breach can swap the core with water. This water is then designed to flow within the reactor building continuously, cooling the core and preventing further issues (figure 5.13).

So, the key lessons learnt from TMI-2 focused on the core itself, with new reactor designs taking the potential problems into account, and developing strategies for eradication, or failing that, minimisation. The key outcome for the USA after TMI-2 was an immediate stop to the building of new nuclear reactors, which prevailed until 2007, when work resumed on the reactor at Watts Bar (Watts Bar 2) in Tennessee.

One of the key challenges when operating a reactor is ensuring that the fission process is continuing and has not stopped. This challenge affected Reactor 4 at the Chernobyl facility in Ukraine, which experienced some of the same issues that affected TMI-2.

9.3 Chernobyl—Reactor 4

Chernobyl is a name that is now infamous among the nuclear community. The reactor complex itself is located near the city of Pripyat in Ukraine, near the border with Belarus, providing power to the nearby region and Kiev. The incident at Chernobyl Reactor 4 highlighted that the design of reactors is key to their safe use, and emphasised that a lack of a safety culture can lead to dramatic outcomes.

9.3.1 Reaktor Bolshoy Moschonosty Kanalny (RBMK)-1000

Before we discuss the Chernobyl Reactor 4 incident, we need to outline the reactor type used, as it is non-standard in design, and in retrospect came with a design flaw. The reactors at Chernobyl were all Reaktor Bolshoy Moschonosty Kanalny (RBMK)-1000s, with an output of ~1000 MWe (3200 MWt), and in total there were four that were operational, with a further two under construction [7]. The RBMK design could be considered a hybrid of an AGR, with graphite moderation, and a PWR, with light water as the coolant. In this design, the fuel was loaded into Zircaloy tubes and placed into a perforated graphite block, which also allowed the control pods to be inserted. Water was then passed through these Zircaloy tubes, cooling the fuel, and becoming steam, which was then used to drive turbines [11, 12].

This design is different from that for a traditional PWR, which embeds the core within a water reservoir that works as both as coolant and moderator. There are advantages and disadvantage to each design, and one advantage of the RBMK-1000 was that it could be used to generate power and breed at the same time, thus utilising unenriched fuel. A second advantage was its ability to refuel online without a reactor shutdown, increasing the operational economic efficiency. An illustrative schematic describing operation is shown in figure 9.8.

However, while there were multiple advantages to the design, there was a design flaw that later became part of the problem at Chernobyl. The reactor had a large positive void coefficient. In chapter 5, we discussed the effects of the void coefficient in the operation of a reactor, with a negative void coefficient being inherently safer.

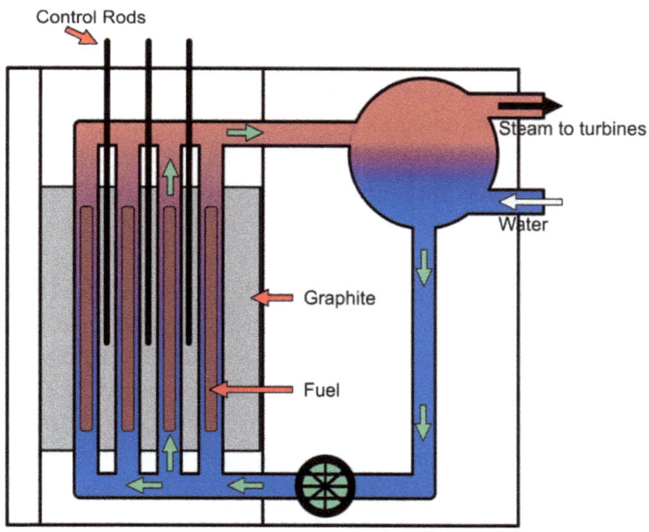

Figure 9.8. Schematic of a portion of the core within a RBMK-1000, highlighting the graphite moderator, fuel and cooling systems.

The RBMK design at that time had a large positive coefficient, and while this was not dangerous in itself, it would not help if the reactor became unstable. The positivity arose from the use of graphite as a moderator, which offset the negativity of the cooling water, and unfortunately it also meant that if there was no water in the system, the fission process could accelerate when the control rods were not inserted quickly enough.

One of the concerns about the design was based around a power failure, and the switchover to diesel backup generators. The question raised was, 'could the turbines continue to turn after an emergency shutdown until the backup generators switched in?' In essence, how long would the switch take, and what would the impact on the core be, i.e. could it continue to get to a safe state? In the majority of reactor designs, the control rods are controlled with their default state in the reactor, in order for the reactor to operate the controls must effectively be forced out of the core. This was the same in the RBMK design, and as with a lot of reactors, the control rods came in from above the core, using gravity to assist, if necessary.

To test the situation where the power switched over to diesel back-up generators, Chernobyl Reactor 4 was selected for a test, and this was set for 25 April 1986.

9.3.2 Timeline of incident

The timeline reported here is taken from multiple sources, which sometimes contradict each other. However, the timeline is a generally accepted version of events [11–14].

25 April 1986

The test was planned such that the power of the reactor would be reduced down to ~25% of maximum output, when the steam turbines would be isolated from the main power and the performance monitored while the diesel generators kicked in. In the ideal case, the turbines would still be spinning when the generators started, ensuring that cooling water was flowing. The reduction in power was started at **01:00** and by **14:00** the power had reduced to 50% of capacity. However, the city of Kiev at this time requested that power be kept running due to a shortage of power. At **23:10** the power reduction was restarted and continued, across a shift change in the plant.

26 April 1986

00:10. The power output within the reactor had dropped to 720 MWt, which was the level set for the test. However, the output kept falling until it reached a level of 30 MWt at **00:28**, a level too low for the test. To allow the test to continue, the control rods were withdrawn from the core—there are multiple accounts of why and how this happened, but suffice to say, the control rods were withdrawn. In total, there were 211 control rods, some inserted automatically and others manually, but now only a few remained within the core. As expected, the reactor began to increase in power output.

01:05. The reactor achieved 200 MWt and the test began, with extra circulating pumps being put into operation. However, this resulted in there being too much flow, and two pumps were shut down, as the water was slowing the fission process down.

01:23:04. The test began with the steam isolated, and the plan was that by **01:23:43** the diesel generators would have started and the test would have been completed. However, something far more dramatic occurred.

Once the turbine was turned off, the water in the reactor began to boil uncontrollably, and the reactor became supercritical. At **01:23:36** a SCRAM was initiated, which should have dropped all the control rods in and begun to quench the fission. However, this process made the situation worse, as the control rods had a graphite tip to aid insertion, and as a consequence during the initial insertion the control rods increased the amount of fission—why?

Control rods slow the reactor down by absorbing neutrons, often based on high neutron absorbers, such as Cd, B or Gd. In the RBMK design the graphite tip, coupled with the control rod displacing water, initially increased output, and this was not helped by the slow speed of insertion at 0.4 ms^{-1}, a speed that meant that it would have taken a significant time for the control rods to fully insert. At Chernobyl, when the control rods were inserted, it was discovered that they only went in partially, and when attempts were made to increase the degree of insertion, these proved fruitless.

Now the reactor core became highly unstable and began to feedback on itself, rapidly rising in temperature and heat output, until at **01:24** explosions were heard. The first explosion was probably due to steam surrounding the fuel coming into contact with water in the cooling system. This explosion now changed the situation, and if it was possible, made it worse: there was now extra oxygen in the system. As with TMI-2, the risk of a hydrogen explosion was present from the oxidation of the Zr-based cladding, which in this case occurred, giving rise to a second explosion that exposed the core, which was on fire. The explosion had vaporised part of the core, which flew up to high altitudes. A photograph showing the core after the fire was extinguished is shown in figure 9.9.

Figure 9.9. The exposed core at Chernobyl-4 [15].

9.3.3 Aftermath

Within minutes of the incident, local firefighters were present, trying to put the fires out using a range of devices, and the fires outside the core were extinguished by the early morning. However, the fire within the core was not extinguished until 10 May, with the expectation that a significant proportion of the core had burnt.

The local population were not immediately evacuated away from the area, and remained for some time, 24 h after the incident. Removal was initially for three days, but this has since become permanent. The effects on the population are still being examined 30 years after the event.

As a sign of the times, the international community did not initially become aware of the issue at Chernobyl. It only became apparent when radiation detection alarms were going off near other nuclear power plants, and it was then officially acknowledged on 28 April.

The radioactive material dispersed into the atmosphere and travelled for some distance, contaminating wide areas with ^{137}Cs and ^{90}Sr, as far away as North Wales in the UK. The local area now has an exclusion zone, and the likelihood is that it will remain that way for some time, if not permanently. As for the reactor itself, it has now been cocooned within a concrete sarcophagus that allows examination of the interior.

9.3.4 What caused the incident?

There are multiple opinions over what caused the accident, whether it was operator error, a reactor design flaw, or other causes. What can be worked out, however, is that the reactor experienced a coupling of a LOCA with positive feedback from the graphite moderator. These led to the reactor becoming unstable, ultimately leading to the explosion, but when did the reactor start to become unstable?

The key step to answering this question is the drop in power to 30 MWt—what could have caused this rapid drop? In chapter 4, we discussed fission product formation, and highlighted one in particular, ^{135}I, which undergoes beta-decay, with a half-life of 6.52 h, to ^{135}Xe, which has a very high neutron absorption cross-section of 2650000 barns. This builds up naturally in a reactor during operation. reaching secular equilibrium.

What occurs in a normally operating reactor is that as the ^{135}Xe forms, it absorbs a neutron to form ^{136}Xe, or undergoes decay to ^{135}Cs (half-life of 9.52 h), both of which have relatively low neutron absorption cross-sections. As the reactor operates, this formation equilibrates and is no longer a major operational problem. However, if neutron production drops, through a reduction in power or through a shutdown, the transmutation of ^{135}Xe no longer continues, and instead it rises from the decay of ^{135}I. This constitutes a problem, as now, when the power increases, a significant portion of the neutron flux, at least initially, will be absorbed by the ^{135}Xe, and not used to induce fission. Once the ^{135}Xe begins to reduce in concentration, the neutrons that would previously be absorbed are now available for fission, and an increase in power output can arise. A similar process is car wheel spinning from too much power when waiting at the traffic lights; eventually the wheels grip and there is a massive burst of power, which can be uncontrollable. This uncontrollability is what is likely to have occurred at Chernobyl-4.

As the power was being increased from 30 MWt to 200 MWt, a significant portion of the neutron flux would have been transmuting the ^{135}Xe to ^{136}Xe, and as the level reduced, the power would have spiked.

9.3.5 Key outcomes from the incident

From a materials perspective, the biggest change came with the redesign of the RBMK-1000, with the focus now being on reducing the highly positive void coefficient. This was achieved in multiple ways, but concentrated on the following areas.

(i) An increase in the number of control rods, with an increase in the speed of insertion.
(ii) Additional absorbers to inhibit operation at low power, i.e. stop the reactor if the power output is too low, effectively putting the handbrake on.
(iii) A slight increase in enrichment to accommodate the changes in control design.

However, by the far the biggest outcome from Chernobyl-4 was the acceptance of the risks in the design. As a result, no further RBMK-1000s were built, although a significant number, 11, remain in operation today [7].

In a manner similar to the Chernobyl-4 incident, what happened at Fukushima sent shockwaves through the nuclear community, and arose from a similar situation to that at TMI-2, a loss of coolant.

9.4 Fukushima Daiichi

The Fukushima Daiichi reactor complex is located near the towns of Okuma and Futaba in the Fukushima prefecture, Japan. Nearby (~12 km) is the Fukushima Daini reactor complex, comprising a total of four reactors.

The incident at Fukushima, whilst severe and wide-ranging, may not offer many lessons that can be learnt for future nuclear materials development. In many ways, once started, it was grimly predictable. However, given the nature of the incident, it was difficult to resolve. Before we discuss what happened, what design were the reactors at Fukushima Daiichi?

9.4.1 Boiling water reactor (BWR)

The reactors, six in total, at Fukushima Daiichi were of the boiling water type, based on a design by General Electric, and were built in stages from 1967 until the final one reached criticality in 1979. The reactors' output ranged from 460 MWe for Reactor 1, through 784 MWe for Reactors 2–5, and up to 1100 MWe for Reactor 6 [7]. A schematic of a BWR core, and an image of the reactor complex at Fukushima Daiichi, are shown in figure 9.10 [16]. The reactors were operated by the Tokyo Electric Power Company (TEPCO).

The BWR design differs from that of a PWR in one major way; whereas in a PWR, the primary cooling circuit heats a secondary cooling circuit, which in turn drives the turbines, in a BWR reactor the coolant passes through the core, becomes steam, and drives the turbines directly. As with most things, there are advantages and disadvantages to each design, with cost being a factor. It is sufficient to say that whilst the design played a role in the incident at Fukushima, it was not the major contributing factor.

In a manner similar to TMI-2, but unfortunately this time not fictional, the incident was foreshadowed, in this case by a report published by TEPCO that indicated that they would be unable to withstand a tsunami similar to that caused by

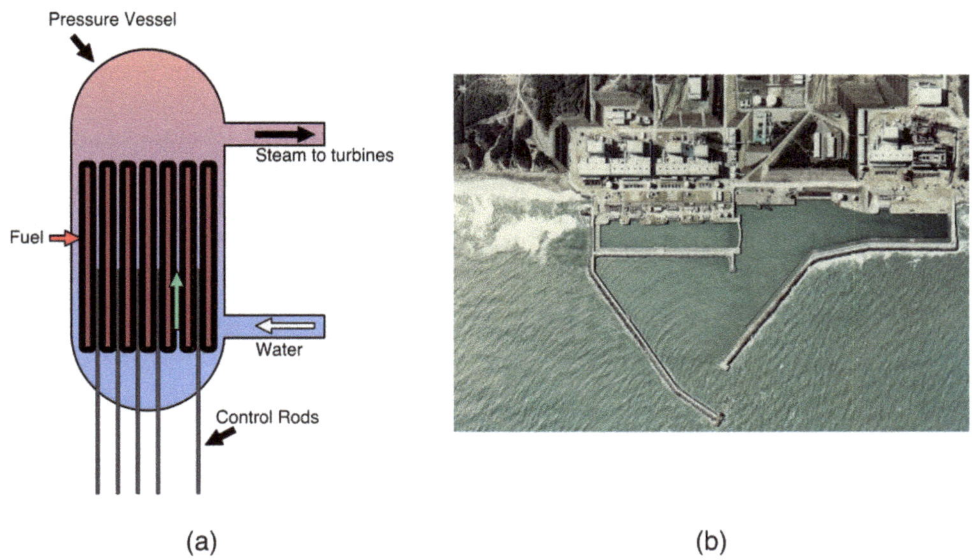

Figure 9.10. (a) Illustrative schematic of a BWR core. (b) The site at Fukushima Daiichi, showing the six reactor locations . Copyright © National Land Image Information (Color Aerial Photographs), Ministry of Land, Infrastructure, Transport and Tourism.

the historical Jogan earthquake (magnitude 8.4–9, 9 July 869). The expected height of such a tsunami would be ~9 m. The report was submitted on 7 March 2011 [17].

9.4.2 Timeline of incident

The timeline reported here is based on multiple sources, which sometimes contradict each other. However, the timeline is a generally accepted version of events. Given the nature of the publicity and complexity surrounding the incident, the timeline focuses on those incidents that impacted directly on the reactors, and it is not a complete breakdown of the events [17, 18].

11 March 2011

14:46. A magnitude 9 earthquake occurred off Honshu at a depth of 24 km. The operational reactors (1, 2 and 3) immediately shut down; Reactors 5 and 6 were undergoing maintenance and were already shut down; and Reactor 4 had been defueled in 2010. At **14:52** the emergency cooling systems in Reactor 1 switched on, but at **15:03** these were switched off.

15:27. The first tsunami struck the complex, it was the first of three, and the waves eventually, at **15:46**, reached a height of 14 m, well above the 9 m predicted for a Jogan-type earthquake. This wave completely inundated the site, causing a loss of power and dousing the backup diesel generators. As a result, there was no more active pumping within the reactors and they were relying on natural cooling.

18:00. It was discovered that within Reactor 1 the water level had fallen and was beginning to reach the top of the fuel, and at **18:18** the pumping within Reactor 1 restarted. However, by **19:30** the fuel had boiled the water away, and it was now exposed—a LOCA had begun, similar to that at TMI-2.

12 March 2011

04:15. The fuel in Reactor 3 was now exposed, as the battery power operating the pumps failed, and there were now two LOCAs running in parallel.

05:50. Fresh cooling water was added to Reactor 1, however, this was insufficient, as by **06:50** the core had melted (this was not known at the time).

10:09. Venting of vapour from Reactor 1 to reduce the pressure began, which was followed by water injection, which stopped at **14:50**. At **15:36** an explosion occurred in the outer building of Reactor 1, and at **19:00** seawater was injected into the core, while steam continued to be released out of the core.

13 March 2011

02:42. The coolant injection in Reactor 3 stopped, and the water levels dropped, leading to core damage (**09:00**). Reactors 1 and 3 continued to be flooded with water containing boric acid, however, the pressure continued to remain high. Meanwhile, Reactor 2 had a lower than normal volume of water, but was considered safe.

14 March 2011

11:01. The building around Reactor 3 exploded, the explosion damaged the cooling water supply for Reactor 2, which failed at **13:15**, and the water began to drop. Core damage began late in the evening. At **15:00** the core in Reactor 3 melted and dropped to the bottom of the pressure vessel.

15 March 2011
06:00. An explosion occurred above the spent fuel pond and the core of Reactor 4. At **11:00** a second explosion occurred in Reactor 3. At **20:00** the core in Reactor 2 melted and dropped to the bottom of the pressure vessel.

This does complete the timeline of events, but for the purposes of this discussion, these events are sufficient to describe what occurred in the reactors; in essence, three LOCAs were running in parallel.

9.4.3 Aftermath

The effects of the incident were wide-ranging, with a large evacuation of people and an exclusion zone that still exists today. The ongoing incident at Fukushima Daiichi was headline news around the world, and it still continues to draw substantial interest whenever it is reported, with the explosions appearing on TV news, as shown in figure 9.11. The area surrounding the facility is severely contaminated, and efforts to minimise the contamination of the biosphere are ongoing.

What is sometimes overlooked is that the reactors shut down immediately after the Earthquake, it was the tsunami that inflicted the damage to the support systems. As a direct consequence, all reactors in Japan were shut down, and worldwide accident readiness and flooding plans were examined to address any concerns about similar incidents.

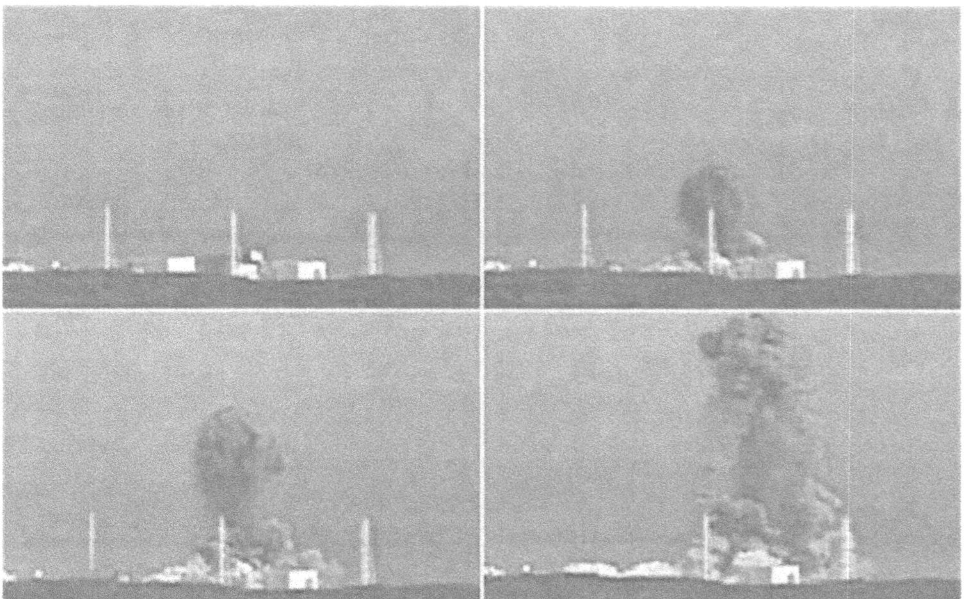

Figure 9.11. Explosion at the Fukushima power plant. Reproduced with permission from [19].

9.4.4 What caused the incident?

Put simply, Fukushima Daiichi was a case of three parallel LOCAs, with there being no means to alleviate the problem because of the nature of the damage surrounding the facility. Once it became apparent that the power had not failed, more specifically it had wiped out, the process occurring within the cores could be predicted grimly. There were some issues raised in the investigation with regards to accident management, but given the nature of the event and the state of the surrounding infrastructure, these could be expected, and understood.

9.4.5 Key outcomes from the incident

In many ways, the outcomes were predominantly improvements in incident management. The incident itself, in an unfortunate manner, did not give rise to major material design changes, as it closely matched TMI-2, this time within a BWR (TMI-2 was a PWR). The material issues, however, were the same, i.e. rapid oxidation of fuel/cladding giving rise to hydrogen gas and melting of the core.

One area where research was reinvigorated and expanded was the development of fuel that is essentially accident-tolerant; not fuel that is immune to effects, but fuel that can tolerate the conditions during a LOCA for significantly longer than the fuels currently used. The research is focusing on the coating of Zr-based cladding to enhance its resistance to steam. This work links in with the design modifications made for GenIII/III+ reactors in dealing with such an incident, and the lessons learnt from TMI-2.

9.5 How do the incidents compare?

The IAEA has a classification to describe the severity of an incident, ranging from 1–7, depending on the severity [20]. The scale is shown in figure 9.12. As can be seen the two with the highest degree of classification are Chernobyl Reactor-4 and

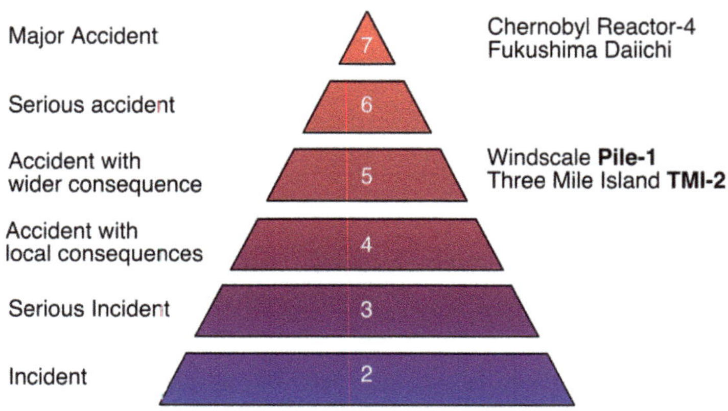

Figure 9.12. The IAEA classification of nuclear and radiological events, showing the incidents at Windscale, Three Mile Island, Chernobyl and Fukushima.

Fukushima Daiichi, i.e. the two with the highest impact. There is disagreement of whether Fukushima Daiichi should be classified as 7, the same as Chernobyl, or higher as the incident was larger.

From a materials viewpoint, TMI-2 and Windscale Pile 1 may have had an equal impact, as they both highlighted the potential risks in reactor operation and the changes needed to minimise their reoccurrence. However, what can be seen is that when an incident happens at a nuclear reactor, the outcome can be dramatic and severe.

The design of future reactors now includes a change in perspective from that seen in the early generations of reactors. Where previously safety assessments were based on the premise of what was likely to go wrong, they are now based on the premise of what could go wrong, and what the outcome would be.

What remains key, however, is that if the materials within a reactor are unsuitable, the chances of a major incident increase dramatically, and a good understanding of how materials behave in a reactor is paramount for their safe use.

References

[1] Sellafield Sites

[2] Wigner E P 1946 *J. Appl.Phys.* **17** 857–63

[3] Environment Agency Wigner energy in irradiated graphite and post-closure safety *R&D Technical Report* P3-80/TR (Environment Agency)

[4] Windscale A L *Windscale* 1957 *Anatomy of a Nuclear Accident* 3rd ed (London: Macmillan)

[5] Wakeford R 2007 *J. Radiol. Prot.* **27** 211–5

[6] UKAEA 1989 A revised transcript of the proceedings of the board of enquiry into the fire at Windscale Pile No1, October 1957 (UKAEA)

[7] IAEA *Nuclear Power Reactors in World*

[8] Report of the President's Commission on the Accident at Three Mile Island

[9] Nuclear Regulatory Commission *Backgrounder on the Three Mile Island Accident* www.nrc.gov

[10] Perrow C *Normal Accidents: Living with High-Risk Technologies* (New York: Basic Books)

[11] IAEA 2005 Accident analysis for nuclear power plants with graphite moderated boiling water RBMK reactor *Safety Reports Series* 43 (Vienna: IAEA)

[12] Lederman L 1996 *IAEA Bulletin* 1996 10–17

[13] IAEA 1991 Safety culture *Safety Reports Series* 75-INSAG-4

[14] Nuclear Regulatory Commission 1987 Report on the accident at the Chernobyl nuclear power station NUREG-1250

[15] The Associated Press

[16] Ministry of Land, Infrastructure, Transport and Tourism, Japan

[17] Tokyo Electric Power Company 2011 *Fukushima Nuclear Accident Investigation Report (Interim Report—Supplementary Volume)*

[18] National Diet of Japan 2012 *The Fukushima Nuclear Accident Independent Investigation Commission*

[19] Nippon Television Company, Japan

[20] IAEA *International Nuclear and Radiological Event Scale*

www.ingramcontent.com/pod-product-compliance
Ingram Content Group UK Ltd.
Pitfield, Milton Keynes, MK11 3LW, UK
UKHW051853100926
8556IPUK00005B/601